W9-BXB-604
Gordon College
Wenham, Mass. 01984

SCIENCE

An American Bicentennial View

Commentaries from a Series of Academy Forums

NATIONAL ACADEMY
OF SCIENCES

Washington, D.C. 1977

Library of Congress Catalog Card Number 77-81904

International Standard Book Number 0-309-02630-X

Available from:

Printing and Publishing Office
National Academy of Sciences
2101 Constitution Avenue, N.W.
Washington, D.C. 20418

Printed in the United States of America

The Academy Forum

Alvin M. Weinberg, Director, Institute for Energy Analysis, Oak Ridge

David A. Hamburg, President, Institute of Medicine, *ex officio*

Courtland D. Perkins, President, National Academy of Engineering, *ex officio*

FORUM STAFF

Robert R. White, *Director*
M. Virginia Davis, *Staff Associate*
B. Saunders Turvene, *Editor*
Marcie S. Lofgren, *Administrative Secretary*

Contents

Foreword

This selection of commentaries was drawn from a series of Academy Forums held in 1975-76 in celebration of the Bicentennial. It contains a wide range of feelings and thoughts about the past and future uses of science and technology--hope, despair, anger, pride, and frustration--as well as some new perspectives and occasional profundity. It says much about citizens and science, scientists and society.

Although it is impossible to recognize all those who translate a concept into an Academy Forum, we wish to make grateful acknowledgment:

To the Council of the National Academy of Sciences, who first suggested the series; the General Advisory Committee of the Forum, who conceived it; and to Philip Morrison, who nurtured it.

To all who, in the course of this series, shared their reactions and responses to a set of grand propositions with care and candor.

To these sources of support that made possible this publication as well as the series from which it was developed: Cabot Foundation, Inc.; Exxon Corporation; General Foods Corporation; International Business Machines Corporation; National Academy of Sciences; Rockwell International; Alfred P. Sloan Foundation; Union Carbide Corporation; and Union Oil of California.

And to B. Saunders Turvene, who saw the shape of this volume in a mass of material and prepared it with skill and judgment.

Robert R. White
Director

SCIENCE

An American Bicentennial View

Commentaries from a Series of Academy Forums

Public Images, Private Reflections: *An Introduction*

PHILIP MORRISON
Institute Professor, Department of Physics, Massachusetts Institute of Technology

Ours is a new world and a young nation. But it is not that new a world: men have hunted and women have gleaned here for ten thousand years and more. It is not so young a nation either: our institutions were revolutionary two centuries ago, and we marked that anniversary in 1976, as all Americans must know.

The Academy Forum, whose whole purpose is the public airing of issues growing out of the discoveries of science and the uses of technology, was caught up in some need to look back and to leave a record of what we saw. It was no novelty for the General Advisory Committee of the Forum to plan a discussion about matters of deep concern and serious controversy, to prepare a theater of advocacy in the House of Sciences. We embarked on a plan that was modest in scale but challenging in content, a look more philosophical than usual. We undertook a probing which would see American science and technology not in their present context alone, not in specific tasks with the disciplines set in line, but rather as streams--clear or muddy, we would not say--which join to form the current of history. We would put big questions, not hoping to answer them in the brief space of half-day sessions and the modest diversity of a half-dozen speakers, but hoping to ask them well and to commend them to us all. For answers they demand. It is, of course, the answering and not the answer that is our real goal. Large questions require large-minded effort, but they usually do not gain crisp little answers.

This is our record, then, drawn from a series of three Academy Forums in a Bicentennial context. It is a thoughtfully chosen sampling of responses to questions posed in the following impressions of each session. The Forum is an ongoing process within the National Academy of Sciences;

we hope that our readers will formulate their own responses and that the more energetic will communicate them to us.

FORUM I SCIENTIFIC THEORIES AND SOCIAL VALUES

> It can be argued that only twice in our history have the structure and results of science had a direct effect on our values, determining the wider texture of life not by virtue of any economic or military consequences, but within the domain of attitudes and world view. The first was that Newtonian world of mechanism, which via Locke, Montesquieu, and the Enlightenment set in motion ideas that are included in the Constitution by which we still live. The second was social Darwinism, which flowed from the self-serving interpretations given the theory of natural selection. To what extent can the claim of a profound impact for these earlier scientific breakthroughs be sustained? Can we see in modern science today--in Einstein, Bohr, in the central dogma of DNA, or elsewhere--that a similar impact has been realized, or is potential? Or is it the *material* results of science and technology that alone shape our world? Where has new knowledge through its structure of theories taken us, and where will it lead us?

So read announcements of the first Bicentennial Forum. But this proved to be no era for theorists. Our historians argued, quite reasonably, that the Enlightenment sought order and law long before Newton explained the orbit of the moon from the fall of the Kentish apple. Newton became the hero of the philosophers because he confirmed in a distinct realm of thought what they had already argued for marketplace, statecraft, and religion. Nor did the glorification of the entrepreneur, of the self-made man, of the necessary goodness of the unfettered competitor, really wait upon Darwin's careful analogies between dog breeding and the Galapagos finches. Rather, eager advocates seized upon the competitive element within Darwin's much more general view of selection on the long time scale of species change and applied it forthwith to justify by its brilliance their pet apology for a system that built fortunes and wretchedness at one and the same time. Nor in the Forum discussion did theory remain central: the problems of the world kept creeping in.

Still, science is not only a dramatic power for modifying the world out there. It modifies even more strongly, one can maintain, the inner landscape of human thought. This green footstool of God round which the ardent sun obediently circled was set adrift forever by Copernicus and the astronomers. It took longer to put our lives squarely into time. Newton was willing to accept Creation as in 4004 B.C., more or less. Not until the geologists and paleontologists gathered around Darwin to proclaim the antiquity of the land and its life did we begin to see our place in the billions of years of galactic history. It was Rutherford who, after the turn of the twentieth century, had to explain to the old and crotchety Kelvin that the sun was indeed billions of years old and that nuclear decay could prove it.

Add to the span of time the twentieth century's own triumph: the galaxies, just as far away as time past is old, and you have the outflowing universe we now inhabit. It is not easy to avoid the conclusion that the secularism that dominates our era--and its dialectical counterpart, the return to a spangled and diverse religiosity and a moony search for revealed meaning--is the consequence of a universe so lacking in center.

These notions do not yet ride the surface of the public mind. Maybe they are there in the depths: the unity of life in its helical inwardness, the speech of the chimpanzees, the life and death of oceans, and a hundred remarkable results and conjectures. The view of the blue earth from space is the one image out of science and technology that has clearly caught the popular fancy. But what has been grasped is that these ways of thought and action have power. In a world beset and frustrated by the close retention of power within the small circle of the powerful, power fed by power, not much voice is spent talking of the viewpoint that in some way commands our conceptual vista. The resonance is lost in the noise.

When Henry Adams wrote, that was not so. He saw the power in the dynamo and the threat in the universal crisis, but he spoke of the idea behind that spinning icon. Thermodynamics was then squarely in the world of the imaginative, but in 1976 it could attract few of the reflective. There is too wide a gulf between the adepts and the thoughtful, a gulf into which much will fall as long as it remains unbridged.

FORUM II THE CITIZEN AND THE EXPERT

> We are aware of the tensions between the specialist, the expert, and the man in the street, the external critic, the skeptic, the citizen who forms his own opinions. Where does the truth lie today? How far can we rely upon experts? Who trusts them? Do our institutions--from the three branches of the federal government to the structure of the National Academy of Sciences itself--contain room for expert opinion and citizen decisions?

In this Forum there was ample coupling: the experts and the public had concrete grievances, by no means one-sided. They are put quite strikingly in the excerpts; no one's case lacks a fluent advocate.

It seems a little strange that this division, too, has become so sharp. For if I am not mistaken, the Republic has always and forever been governed--under ballot and law--by the experts. They were, as they remain, experts of two disciplines, lawyers and generals. Lawyers write the law, administer it, judge its execution. Presidents--if they are not generals--are lawyers. So are the congressmen, the judges, and down the ranks, to people below the point where political power rests. Those men and women--women in office are also mainly lawyers, though not yet generals--share a formal training, a jargon, a discipline, a system of credentials, and a community. Swift is ironic enough about lawyers; so was Mr. Dooley, for that matter. Nowadays they are not far from heroes. They sue and threaten on behalf of smokers and patients, auto buyers or Indian tribes. The generals are not heroes right now, but they certainly were up to one lost war ago.

Why now do we fear and mistrust the expert engineer or scientist? Their power is less than that of the management, their bosses--lawyers or generals with a few exceptions. There may be seen some reasons. Lawyers deal in process, not in content. They are seen learning from witnesses, from clients, asking for explanations, weighing the answers of the contestants. Scientists don't do that: they consult their peers, or the literature, or worst of all their instruments and their samples or the animal colony. If they ask questions of people they make a federal questionnaire out of it and simmer the results with various statistical spices.

There is good news to report in this post-Bicentennial period. More than one small city, ambivalent host to a great university or two, is now wrestling with the legal problems around the possible health hazards of laboratory experiments with recombinant DNA. So far the key burden of proposing regulations has been given at least twice to citizens' committees set up for the task. One university person trained in science, though not in nucleic acids, has been on the committee, a full member, but no more than that. In the cases I have described the consequences were extraordinarily Jeffersonian. The members met seriously, they asked, they read, they argued, they decided. Their proposals were judicious, informed, free from cant or arrogance, to the point, and unifying. These public representatives neither abdicated to the experts nor neglected them. They themselves became informed, responsible, and decisive. It is a parable for our future. For the one guiding light of science and of the technology it nourishes has always been the public welcome it offers to all those who would understand.

Science sets no aristocratic condition of any kind: study the literature and think about the issues still uncertain, and you--anyone--can understand. No barrier save that of time and effort stands between expert and citizen. Both know how that barrier can be crossed; they must mutually try.

Perhaps here the lawyer and the general have had an easier task. Somehow the experiences of daily life--misunderstandings, dishonesty, confusion of memory, conflict of interest, the judgment of trustworthiness--make the lawyer's work, his assumptions, his experiences, familiar. The general, too, at least in so far as he deals with troops and with enemies, has a link to Everyman. For Everyman served at Fort Dix and even went overseas once himself, or anyway his niece or uncle did. Armies are not as strange as are chromatographs and chi-squared tests. There is a lesson here written very plain. We need to bring the public into the lab and up to the desk, not only along the path of correct conclusions but, even more importantly, to the tangled ways of error. Inclusiveness and not authority is the implied requirement of science done in public.

FORUM III THE USE OF KNOWLEDGE: FRONTIER EXPANSION OR INWARD DEVELOPMENT

> The American Frontier spread westward to the sea. Our overseas expansion has had a limited but real effect. Yet today we look inward to the problems of the city and the despoiled waterways more than to spatial boundaries to confront our deepest problems. So it may be in science itself: it is new knowledge and new techniques in new fields that mark new frontiers. Yet the Academy finds itself involved more and more with problems of the management of the established ground, and rather less in seeking new lands of knowledge. Is this right? Is it inevitable? Is it a result of a temporary imbalance? How can we judge what effort and concern we ought to spend looking outward or inward? Who can judge? Do we all stand equal before such a judgment?

In this Forum, most participants, although not all, found a comfortable middle road. If the convenient aerosols threaten the ozone layer, well it certainly took research on the upper atmosphere, more or less free from programmatic hopes, to find it out. If you would argue--some did--that we do not need and should rather shun all the easy conveniences and public comfort and health of a "developed" nation, there still remains the chance that nitrate fertilizers might do the ozone in. And even if you call for zero growth and let the overly large generations go hungry, you will find it hard to argue that what you don't know won't hurt you. Laissez-faire is unpopular among the ill in every society, even if nature does usually heal.

On the other hand, it is plain that American knowledge, like American wealth, is skewed in its distribution. One percent of our citizens own a quarter of our wealth, and a similar peak applies to science. The task on the side of practice definitely requires concern for the unexamined consequences of the embodied past of our development; on the side of theory, it seems to require letting in the rest. We look to public education in science--demystification is the trendy phrase--and a new collegial hope for many new scientists and engineers, those whose congenital roots, in ethnic origin, social class, geographic location, or extra chromosome endowment, have too often kept them out of our councils. The frontiers will be attacked, the battle well-manned and well-womaned, with plenty of force left for interior safety, if we enlarge the recruitment for thesis defense and for academics. The enlargement of science in every way, not just in the scale of grants, might help answer this question as it clearly would help with all the others.

Let the century flow. Whether the nations will themselves survive the present coincidence of overconcentrated powers, both the explosive power and the economic-political power, is not easy to say. But if they do, then the Tercentennial will, I am sure, be welcomed by an Academy and a world of scientists which will take a much wider cut through humankind than the anniversary we now imperfectly and restively celebrate after two hundred years of the Republic and one century of its National Academy of Sciences.

FORUM I

Scientific Theories and Social Values

It can be argued that only twice in our history have the structure and results of science had a direct effect on our values, determining the wider texture of life not by virtue of any economic or military consequences, but within the domain of attitudes and world view. The first was that Newtonian world of mechanism, which via Locke, Montesquieu, and the Enlightenment set in motion ideas that are included in the Constitution by which we still live. The second was social Darwinism, which flowed from the self-serving interpretations given the theory of natural selection. To what extent can the claim of a profound impact for these earlier scientific breakthroughs be sustained? Can we see in modern science today—in Einstein, Bohr, in the central dogma of DNA, or elsewhere—that a similar impact has been realized, or is potential? Or is it the *material* results of science and technology that alone shape our world? Where has new knowledge through its structure of theories taken us, and where will it lead us?

ROBERT McC. ADAMS

Harold H. Swift Distinguished Service Professor, The Oriental Institute

Our concern is with certain grand themes that affect science as it is embedded in American life, with the lines of influence running in reciprocal directions. The kinds of issues that confront us within the scientific community today are, frankly, very similar to those that confront the society at large. Important among them, for example, is the challenge of accommodating an enterprise that continues to increase in scale and complexity despite a steady or even declining flow of the resources needed to sustain it. Do we need to reverse values once attached to cumulative growth, or can we merely redirect them into an appreciation of subtler, less expensive refinements? How can we approach a steady state without unduly restricting opportunities for those just entering the system, and without foreclosing untried but potentially fruitful lines of inquiry? It would be misleading to ask such questions, let alone to try to answer them, for science alone.

A second deep quandary within the scientific community, as well as outside it, involves the challenge of maintaining a consistent orientation and sense of coherence in an atmosphere of social fragmentation and institutional flux. Probably to be included under this rubric is advancing specialization, since many barriers to intercommunication are direct outgrowths of it.

Like the societies that support it, science may in some respects be viewed as a pluralistic ethical and value system. From that standpoint, however, can the standards, presuppositions, and priorities that were intimately involved in the achievement of scientific excellence in the past remain impervious to the questioning of and experimentation with value systems that we see going on all around us? Is it reasonable to regard the scientific value system as capable of cumulative growth and refinement but essentially changeless in structure, when the structure of science itself is so obviously changing? The frame of scientific analysis of an earlier generation, for example, tended to be set, with little reflection, at any convenient size. One isolated a particle or an organism or a community, or one worked unselfconsciously within a discipline as if it were a wholly natural, inevitable grouping of postulates, findings, and methodologies. Having drawn a set of essentially arbitrary boundaries around some object of study, scientists felt free to pursue their self-selected, artificially isolated objective to its logical end.

A number of more recent concerns and perceptions that have begun to claim our attention are requiring us to become more critical, flexible and self-conscious about where and how we have placed those boundaries. That is, in a sense, what an ecological approach is all about. It subsumes the organism within the population with which it interacts and the population within its environment, while not in any sense denying that for certain purposes the isolation of the organism remains a powerful analytical tool.

I think that same issue is raised when we are asked: In whose interest is the science that you are conducting? Within what limits is it reasonable to value scientific procedures and discoveries independent of their social impacts? What are the costs as well as benefits of particular segments of scientific activity, and will the benefits accrue to the same groups, communities, and nations as those asked to bear the costs? The urgent, sometimes even angry, asking of these questions requires more than a merely defensive, self-righteous response. It should lead us to a wide-ranging reconsideration of the mechanisms by which we isolate our own research specialties from colleagues in sister disciplines, from problems whose outlines seldom correspond to any disciplinary structure, and from the societies that ultimately must be asked to support our work.*

*The author subsequently elaborated upon the same theme within a more disciplinary framework: "World Picture, Anthropological Frame," *American Anthropologist*, Vol. 79, No. 2 (1977).

CHARLES C. GILLISPIE

Dayton-Stockton Professor of History, Program in History and Philosophy of Science, Princeton University

We live in a society that produces and consumes goods in unprecedented abundance. The technology that makes it possible is based in its progressive aspects on science, and in all its aspects acts upon an exploratory, exploitative, and coercive attitude toward nature. That attitude has been one of the distinguishing features of our modern civilization, differentiating it from all others. Indeed, it is one of the two features, the other being our heritage of Greek mathematical science, with its techniques and its practice of rational proof, that explain why modern science is a product of European culture and not of any other culture ancient or modern. The attitude I speak of first emerged in fully recognizable form in the deeds of the Renaissance engineers and voyagers, and the history of our technics has been continuous since they began imposing their European will upon nature and the world.

Beyond those obvious and fundamental points, however, historians have not yet done very much to refine our understanding of the part that science and technology have played in the historical process. I do not flatter myself with having the insight to penetrate far into that question. A few years ago I did put together a paper about the progressivism of science. Abstracting from questions of economics and industry--I could justify that abstraction, while recognizing that it makes the argument partial--the argument is that science has made itself felt not politically but culturally and has done so through its erosion of tradition and its orientation of sensibility toward the future. Politically, its influence has been neuter, and its role has been to enhance the power of the state while in return drawing advantage for science from the state.

Now I should like to take the occasion of the Forum to say something more, enlarging what we mean by science to include the whole technical culture and enlarging on its relation to the exercise of power in modern history, and perhaps also on the resistance to it, active and latent. For surely it is for reasons of power that politicians are interested at all in technical affairs. Consider, for example, Richard Nixon, that least scientific and most political of men. Why else was his every instinct for pressing forward with the supersonic transport plane, no matter what? If he could have been convinced that basic science contributes to power, as for a time after Sputnik politicians were convinced, rightly or wrongly, why he would have been for that too. Or consider the members of the joint Congressional Committee on Atomic Energy in 1950 and 1951 at the time of its hearings on a crash program for the hydrogen bomb. The Committee was totally uninterested in the moral scruples, the strategic reservations, the estimation of diplomatic consequences, over which Rabi, Oppenheimer, Fermi, Dubridge, and Conant had agonized through endless meetings of the General Advisory Committee on Atomic Energy. All the politicians--including President Truman, no less a politician than Nixon, even if a better man--wanted to know was whether the new bomb would go off, and when told it probably would, they never paused.

A few political scientists are beginning to study the relation of science to war, a relation that goes much further back, however, and that is therefore more systematic than they have yet appreciated. In the French Revolution, for example, the Committee of Public Safety mobilized the scientists to direct the fabrication of armaments and war matériel during several years of military emergency. What is more interesting, the scientists themselves served the regime that had just abolished their Academy, the ancestor of this one, and arrested several of its leaders--notably Condorcet and Lavoisier--and they did so with an enthusiasm later matched by the atomic physicists at Los Alamos. For both sets of scientists, mobilization was the great event in their professional lives, when they proved their civic manhood by way of their science. The subsequent prosperity of physics in this country, though beggaring by comparison everything that went before, was no novelty in principle. Each previous stage in the institutionalization of American science occurred in response to the necessities of war. The National Research Council is the legacy of the First World War; the National Academy of Sciences, of the Civil War.

But war is only the most overt form of exerting power,

and not necessarily the most interesting. If we look at modern history in a very large view, I would hazard the judgment that the most important difference from earlier times is that, since the French Revolution, whole peoples have been the constant preoccupation of governments as they never were before, when the people suffered history rather than took part in it. That is as true of the totalitarian regimes as of the liberal or democratic. In this respect Hitler and Stalin were no less the heirs of the French Revolution than Roosevelt and Churchill. And, clearly, it is the technology of communications, which is based more immediately on science than any other, that made it possible: not that caused it, mind you, for the causes lie deeper in history, but that made it possible, this immediate access to every eye and ear, this detailed information on the actions and often the thoughts of every person. Nothing of the sort was remotely conceivable to either tyrants or popular tribunes in olden times. But the consequences are a two-way street. For the ruler, whether legitimate or not, is also exposed to a scrutiny in detail equally unimaginable to a Louis XIV or a Cromwell.

What this reflection suggests--I venture it as the merest of suggestions, though a good many other examples could be adduced--is that in the conduct of politics, which is the exercise of power over persons, reciprocally the governor over the governed and the governed over the governor--the effect of science is to amplify, sometimes to an enormous and no doubt dangerous degree, what is happening anyway, and usually for other reasons. Thus, science and technology are instances and not causes of the difference between the rationalized economies of Europe and America and the localized economies of Africa and Asia; and instead of abating those differences, technology seems to be augmenting them, and also the awareness of them.

The point I am making first came to mind in connection with a course I teach, where it has to be considered how the supposed internationalism of science can be reconciled with the very different patterns in its development nationally in the major countries. The content may well have been international, but the forms were national and surprisingly persistent. For example, compare the relation of science to the state in Germany in the 1820s to what it was in the 1920s. Humiliated by Napoleon and divided by history, Germany in the earlier period was summoned by her intellectuals to achieve in the realm of science and learning the unity and dignity denied her in political reality. Consider next the Weimar period. In the interval German science and technology had become the wonder of the world. And once again, humiliated by Versailles and divided by faction, Germany was summoned by her intellectuals to sustain in the realm of science and learning the unity and dignity denied her in political reality.

Or consider the United States. In the nineteenth century it was said of us by others and by ourselves that in technical things as elsewhere, our genius was practical, not intellectual, for material improvement, not for scientific theory. Only in World War II did the United States achieve great power status in physics. Even then, however, American physicists brought to bear on the atom none of the basic theories or ideas on which its fission depended but rather the cyclotron, the first scientific instrument of industrial dimensions, and the ability to mobilize large masses of men, money, and equipment to produce a result. I could make similar observations about the French and British modes in science but have said enough to make the point.

The point is that there must be some quality of indifference in science and technology that they can be done so well in ways as different as those of, say, the United States, the Soviet Union, and contemporary China. The same is not true of literature. Let me make what I take to be the same point, though with a very different sort of example. It is the indifference of technology that is producing the pollution that the world has quite suddenly begun to notice. But it will also be technology that will abate it, if it is abated.

Amplification of power, multiplication of problems, indifference to consequences--those are the attributes, among others, that technology brings to polity. There are signs that such indifference is creating doubts about the capacity of society to tolerate the strains. No doubt this Forum is itself one of those signs. Up until now scientists have acted in all their vigor on the unquestioned assumption that any problem that could be solved should be solved, and engineers, on the congruent assumption that anything that could be done should be done. The most interesting attempt I know to question those assumptions analytically, instead of in a gust of emotional revulsion and sentimentality, is Jerome Ravetz's book *Scientific Knowledge and Its Social Problems*.

The problem is a very difficult one, however. To resolve it would require nothing less than modifying the motivation and dynamics of the entire technical culture, and history has very few examples of such a thing happening on purpose in a culture. (In fact, the only example I can think of is the modernization of Japan.) To inhibit or prohibit some line of research because of possible consequences would go against all the grain of science and indeed of scholarship.

How could it even be done? Most of us would agree, I imagine, that the world would be more habitable if there were no nuclear weapons. Yet, reviewing the main steps that led to the explosion at Hiroshima, at what point would one with some authority, even if one could imagine what authority, have prohibited research from going further? After Becquerel's discovery of radioactivity? After Madame Curie's isolation of radium? After Einstein's formulation of the mass-energy relationship? After Rutherford's bombardment of the nucleus? After Lawrence had invented the cyclotron? After Otto Hahn had discovered uranium fission? At what point along the way could anyone have foreseen the bomb? And when Szilard, Wigner, and Teller did see the possibility, would we have had them refrain from pressing the information on the American government when there was every reason to suppose that German physicists had the same information and Germany the same capacity?

Merely asserting that technology has a responsibility to society does not evade such questions. Yet neither does the difficulty of the questions evade the responsibility. And perhaps one of our greatest needs, therefore, is to begin to find some way to answer them within the technical community, to begin to devise some criteria according to which scientists and engineers can ask themselves the question they have never had to answer, which is, "In solving our problems, what else may we be doing?"

* * *

We need to distinguish between the content and the practice of science. The problem with science as a body of knowledge is that, although it can show us how to do all sorts of things, it is incapable of establishing goals. Indeed, throughout history it has achieved its mastery over subject matter at the price of eliminating consideration of goals from its formulations. Science as a system can say that something *can* be done. It never really can tell us that something *should* be done. The problem of getting agreement about goals and practicalities within the whole complex of conflicting interests is in a very large sense a political problem, to the solution of which scientists have never felt called upon to make a contribution professionally. Of course, they have participated as people who need to get themselves supported and to operate in the world, but that is another, subsidiary, matter, concerned with the practice of science as an enterprise.

The difficulty with the summonses being issued to science arises partly from a confusion about the possibilities in science and partly from a failure to consider feasibility in the world. When you can get a measure of agreement in society about a goal, then science and technical capacities often permit you to reach it. Rhetorical comparisons are sometimes struck between our ability to make an atom bomb or to put a man on the moon and our inability to mount, say, a decent system of rail transportation. In the political sense, the former were easy. There was no enormous complex of impeding, conflicting adversary interests preventing the effort. I think that the real problem is in the political process, and that the solution lies in involving the practice of science in that process in a healthy manner rather than in intruding values or morals into the content of science, which is impossible anyway.

CLIFFORD GROBSTEIN

Professor of Biology and Vice-Chancellor for University Relations, University of California, San Diego

Do science and its theories affect human values? The question is a complicated one. We have some difficulty knowing exactly what we mean by "scientific theories." We don't know how broadly to interpret the term or how inclusive to be. We also have difficulty in defining "social values" in a way that enables us to draw diagrams or formulate equations of the interactions between them and scientific theories.

Nonetheless, my own feeling is strong that the two do interact very substantially: They interact primarily because each, even though we abstract them one from the other, represents an aspect of human behavior. Human beings formulate scientific theories and create social values. In the process they may create conflicts between the two; but, over time, hopefully they resolve some of those conflicts.

* * *

The very special thing that science does is to constantly expand our perspective and to force us to see both old things and new things in new ways. Perhaps the best example at the present time comes from the confluence of astronomy and biology in the last half century. This confluence has given rise to what has been referred to as the concept of cosmic progression. This is the notion that we live in an epoch with a very pronounced vectorial component. Whether we use the word or not, we recognize progression by cumulative change with time. This has given to us a sense of place in time. The fact is that today we talk about beginnings of the universe as being of the order of 15 billion years ago, the beginnings of the solar system as 4.5 billion years ago, the origin of life on the earth as 3 billion years ago. We recognize that for some 2 billion years--almost half of the lifetime of the earth--it was populated by very simple organisms only at the level of bacteria and blue-green algae. We know that something over a billion years ago things happened that led to a very dramatic surge in the complexity of life; something of the order of 600 million years ago the record indicates that most of the major forms of life already were present. Then some 3 million years ago there appear in the record prehominids or hominids that led to another astonishing surge of complexity on the surface of the earth.

These evidences of progression are today accepted as scientific facts, or at least as close approximations. They supply a perspective that connects the smallest things, the largest things, the oldest things, the newest things, and puts them together in a way that I feel must necessarily change our system of social values in the next century or two.

* * *

There is an interesting dichotomy, it seems to me, on the question of the degree to which social value influences the course of science. I personally believe that it very much influences the course. I think, however, that it is important to recognize that its effect is in determining questions and not answers. Our times are very important in deciding what sorts of investigations science enters into. It would be more difficult to show that the answers of those investigations are determined by social value. It is the answers that, in fact, are most significant in affecting social value.

* * *

Issues like the Quinlan case should not be resolved by an individual physician, nor do I think that they are best addressed in a court of law. It requires a mechanism of collectivity that brings to bear both the expertise of the technologist and the expertise of the humanist, and we haven't figured out how to do this. Our problem is not just to be sensitive to the need but to be ingenious in designing the appropriate social mechanisms.

The university is an attractive possibility or candidate for carrying out a role in this direction that clearly, in recent years, it has not been performing. There are good reasons why it has not. They relate, of course, to what has been happening to the nature of knowledge and what we all recognize to be the hyperspecialization that has both been successful and a danger. But we ought to be able to overcome that. There is a unique opportunity in the university, given the fact that, unlike in think tanks and federal bureaucracies, there are gathered together in universities virtually all the branches of knowledge. When I say gathered together, I mean they are in proximity. It should be possible to establish more effective communication. There are also in the universities students who are, among other things, potential recruits to fields of knowledge both traditional and new and who, in the very process of their education, require some kind of integration of knowledge that the specialists do not ordinarily feel called upon to perform.

It seems to me, also, that the effort on the part of universities, which is not new but which recently has been intensified, to find ways to confront practical problems out there as phenomena of the real world will have, hopefully, an integrative effect on the university and will lead the various disciplines into more effective communication. In any event, I certainly do support the notion that in attempting to find mechanisms that deal better with our topic, there is a dichotomy that somehow has to be overcome, and that the universities should feel called upon to play a role.

* * *

As a biologist I believe very strongly in the importance of the social sciences, and I also feel that for most of my life I have been concerned with human problems and how to solve them. But there are many ways to solve human problems, and the so-called hard scientists very often believe that they are participating in the solution of human problems. They are just not sure which human problems they are on at a particular time: but most of them, if scratched hard enough, will say that down the road what they are doing is going to contribute to the solution of human problems.

I think the complaint is that at the present time the total institution of science doesn't seem to be reacting as rapidly as it might to social problems that we all recognize are going to require the kind of expertise that scientists have--not to solve them with that expertise alone, but to make their significant contribution. What we are complaining about is that science, as a social institution, does not seem as responsive as we would like it to be at this time, when clearly the problems are multiplying and magnifying far beyond the capability of any of us to feel that we are dealing adequately with them.

But I don't think it does much good just to point the finger at scientists. What we are discussing, it seems to me, is how do we more effectively mobilize science as a knowledge-obtaining or knowledge-generating process so that it is more receptive to the guidance of social values. In operational terms I would say that is what we are talking about, and I think we are all aware that there is some lack of registry between these two realms of human experience.

We would like to improve it, and I think when we get to the subject of mechanism then obviously we are talking about what the political scientist wants to talk about. There I would turn the question around and ask which of the political scientists can put before us a model for the kind of interaction that is necessary, whether it be single person or multi-person, and how it generates answers and how it couples with implementative mechanisms, so that what comes out is what makes us all happy.

RAOUL BOTT

Higgins Professor of Mathematics, Harvard University

When we speak about the effect of theory on practice and on society, it seems to me that the relationship is a very indirect one. By this I mean that really most scientific theory affects the society mainly subliminally. It is a sort of fallout phenomenon. I think the marvelous phrase of McLuhan's, "The medium is the message," possibly overstates matters, but there is a tremendous amount of truth in it. Thus, every time we interact with a machine we do learn a little bit that we are supposed to be machines too.

I find this fallout--especially of the developments in physics in the middle part of this century--disturbing, in fact fear inducing. It seems to me they have quite the opposite effect that the Newtonian rationalism had. I react with great fear to the idea that the world is a giant casino and stand here in sympathy with Einstein, who I believe said, "God does not gamble." It is surely a sort of irony that Einstein stands much more in the end of Newtonian thought for the pinnacle, the final crown of Newtonian thought, than for what is going on today. It is not his theory at all but rather the world of quantum mechanics and statistics that has, I think, a very disquieting fallout effect on us.

In a sense we come here into the realm of second-order effects. In applied mathematics this often happens. When you try to explain something, you do it to various orders. In the first order an explanation works very well. But then after you study the situation more carefully, second-order effects, which you first neglected, become dominant. An example, of course, is, if you walk a platoon over a bridge. Then the oscillations, depending on the step, can get into resonance, and the resonance can then destroy the bridge. Resonance is a second-order effect.

My feeling is that we are right now in a similar situation with science as a whole. In other words: Is science now exerting a second-order effect on society? Is it on the verge of collapsing the bridge? I think we are in great danger.

Because I am not a humanist and therefore cannot speak with eloquence, let me borrow something from Lewis Mumford who, on a similar occasion, spoke to my point in the following way:

> All our current plans for science and megatechniques must now face a reaction that almost no one until recently conceived as a possibility, that the very quantitative success in these areas would offer as great a problem for our own age as ignorance and poverty and physical helplessness in the face of natural forces have done in the past. In the last decade the results of incessant quantification have become too gross and life-threatening to be ignored. Not only is the planetary habitat befouled by mountains of waste and rubbish, by sewage, chemical poisons, nuclear pollutions, but the same choking up of organic activities, the same insidious depletion is taking place in the mind. Even science, the uncontrolled quantification of information and knowledge, has patently increased the output of rubbish, unusable knowledge, and nonsense, trivial information, and the method of coping with this pollution by means of computers, retrieval systems, microfilms, tape recorders only erupts in larger waste piles, since no effective form of control can be established, unless we reduce quantification at its source.

* * *

I have absolutely no emotional bias against probability in the common-sense sense. Of course, if you make an experiment you only know things to such and such an order. I remember distinctly that my first lab experiment in physics was to measure the length of the table. We learned that you had to measure it, and measure it, and measure it, and every time you measured it you got a different answer--it was a very instructive experiment--and then we took a statistical mean of our numbers.

But it is a quite different story to feel that even the ideal core of the science is statistical. I mean, in the uncertainty relationships there is something much "worse" than that going on. I am not quarreling with it. I am not saying that it is wrong. I am just saying that these are, in a sense, fear-inducing discoveries. They are not reassuring to me.

After all, what good is it for me, subjectively speaking, to know that I am so small, or to contemplate all the immense distances in space, or the time scale in which we live. How is that to be translated into the problems of the day? It seems to me that this is at best completely neutral information and, from the point of view of building actual values, I cannot see an iota of value in it. I am all for that knowledge in the sense that it sets bounds and gives us certain insights into the world. But I don't see that it really helps us to live.

FRANK E. MANUEL

Kenan Professor of History, New York University

I have never performed a scientific experiment nor do I identify myself as a proper historian of science. Most of my life has been devoted to the study of the literary, religious, and philosophical texts in which Western culture since the seventeenth century has given expression to its changing social and moral values. What, then, is Saul doing among the Prophets?

My presence here is explainable by three areas of moderate competence and major concern. First, I may help to illustrate in detail the manner in which the new philosophy, now generally called science, took root and was assimilated by the ecclesiastical leaders of society who generally held the office of declarers--*déclarateurs* Jean Jacques Rosseau would have said--of acceptable moral and social values. I want to highlight the fact that in the seventeenth century, when the new science was born--if it didn't start as of that moment in time, it then became significant--there were ensconced and well-established bodies of men, corps, that declared social values and looked upon science as a danger, an intruder.

When the possibility of confrontation was recognized, there were a number of responses, which I can categorize *grosso modo*. By the way, these responses, secularized and transmuted, are still alive. That is why I mention them. One response was a metaphor of the two books--the idea that there was the Book of Nature and the Book of Scripture, and both of them were equal and autonomous manifestations of Providence. Kepler said that the Book of Nature, which was the province of science, showed the finger of God, Galileo said that it showed the hand of God, and Newton talked of the impress of the arm of God. That is an anthropomorphic progression whose significance I haven't been able to fathom. Advertising the neutrality of science was one way to make it acceptable and avoid a major problem, a war of science and theology. Some of the same postures are evident today, the same attempt to present science as simply declarative and nonnormative.

Another response in the seventeenth century that I would like to dwell on and that also has a contemporary parallel, one that interests me very much at the moment, is pansophia, the idea that man couldn't live with these two separate bodies, the religious ecclesiastical declarers of moral values and the new scientists, who, whether they admitted it or not, were really rather rambunctious in promulgating new moral values. This notion of pansophia demanded that there be an integration of the two forces in society, the ecclesiastical and the scientific. I myself think that we are in grave danger unless the repositories of the religious, literary, moral, and philosophical culture and the technical scientists achieve a new way of talking with each other in the university, which I still consider the unique agency of culture in Western society, and all the more significant because of the demise of traditional religious institutions. I could continue to reflect on the accommodation of science and the religious or moral values of the culture through the eighteenth and the nineteenth centuries.

My second justification for being here may appear even odder. It has to do with a long-time study of the history of utopian thought, which in a peculiar way gives voice to a society's far-out ideals. I have a special interest in the roles that science and scientists are allotted in these utopias. The scientist, though he has many times refused to sip from the priestly cup, has not forthrightly entreated that the cup be removed from him.

The third area of concern to me--and I am profoundly troubled by it--is the enfeeblement of the traditional religious articulators of values. As the fateful year 1984 approaches, I am horrified by the prospect that the state and political leaders are increasing their pretensions to determining and setting social values through the legislative and judicial agencies. I have therefore become convinced that men of learning in science and in the moral and religious traditions of the West--and I insist on these traditions as the primary source of Western culture--have a responsibility for laying down the principles of social value without which a culture cannot exist. Their secular formulations may be less definitive than a theology, but they will gain in humanity what they may have lost in divinity.

The religious man as depicted in a stereotype of eighteenth-century thought was extraordinarily fearful and extraordinarily hopeful. I think that if scientists and those who have dealt with the cultural tradition of the West can meet on some common ground, we may be able to increase the hope and diminish the fear. Here again, the only body I can recognize where this may be possible is the university, and the university, as you know, is in disarray.

* * *

In my tradition the question is always more important than the answer. You may remember Gertrude Stein's last words to Alice B. Toklas: "What is the question?" The Talmudists also took the position that the question was more important than the answer.

There are various kinds of answers. One of the Talmudic manners of settling a controversy is to say the acrostic "Teiku," which means that when the prophet Elijah the Tishbiite comes, he will answer all questions and resolve all dilemmas. So, there always is an answer.

* * *

I feel that science is the last hope of mankind. I am terrified of cultist religions and the counterculture. The question is that we are trying to raise is the following: Has science really, as some of the clichés have it, always taken the position that everything that can be investigated should be investigated? I know the trumpeters of science like Bacon and the bell ringers of science like Campanella better than I know the scientists, I will confess. And these trumpeters and bell ringers were very clear in setting limitations to the activities of science. Bacon placed unambiguous restrictions on what scientists should and should not be allowed to do in passages throughout his works. One of his restrictions was that nothing should be undertaken that was against Christian charity, charity meaning love. In other words, there had to be a consideration of consequences.

That part of the Baconian corpus has been little heeded as science has expanded. Bacon also called attention to the fact that certain kinds of scientific investigation might create a mental disequilibrium in the scientists, and he warned against such investigations. I think he was referring to certain Paracelsians.

What I am trying to say is that science is the latest great manifestation of human capacity. But I want it to join with other value-setting groups in mankind because if it doesn't and allows itself to go astray by going off on its own, then disaster may occur.

* * *

The importance I attach to the university is purely utopian. I mean, these are things that I wish for more than hope for. However, looking around the world, I see no other agency--my choice is *faute de mieux*. I am not speaking out of any overflowing optimism. I find this the only institution that allows within a certain order great complexity, particularly the universities since 1945. The new American university, since 1945, is something that has not developed anywhere else in the history of the world. But somewhere in the process we went awry, and I would hope that we might reconstitute the original ideal.

* * *

It is my judgment from the study of some of Newton's utterances, but overwhelmingly from the study of his theological manuscripts (of which there are a great many, some 3,000 folio pages), that he was always looking to reduce the external world and the biblical world as he saw it to simplicity. *Simplicity* was a key word that appears in one paragraph in one of his manuscripts seven times. I think that simplicity was achieved magnificently in the Declaration of Independence. We reduced the real complexity and tragic quality of the world to a very few simple principles. Of course, to some, such a reduction to simplicity might be a cause for mourning rather than celebration. The year 1976 is also the bicentennial of other great events. Crucial for our problem on the relations of science and society, it is the bicentennial of the death of David Hume; it is the bicentennial of Adam Smith's *Wealth of Nations*; and, *memento mori,* it is the bicentennial of the publication of the first volume of Gibbon's *Decline and Fall of the Roman Empire*.

DONALD FLEMING

Jonathan Trumbull Professor of American History, Harvard University

I would like to begin by asking whether science changes social values or simply forwards persuasive rationalizations for social values to which people are already committed. I don't think that there is an absolutely open and shut answer to this, but I must say that my strong inclination is to think that social values have primacy and that normally what is sought is rationalizations framed in contemporary scientific discourse that will seem to lend increased persuasiveness to the social commitments that one has already made.

I do think this is a case in which concrete examples are everything. The so-called social Darwinism is, as a body of prescriptive teaching about the conduct of individuals and the organization of society, absolutely identical with the social philosophy of laissez-faire, keeping the hands of the government off the economy and supposing that individuals by pursuing their own interests will, in the end, somehow knit up their purposes into some shared benefits.

It is perfectly clear that laissez-faire was solidly entrenched as the dominant social and political philosophy, certainly of the Anglo-American world, as early as the 1840s: the social values that appear in social Darwinism were in place. Nobody, of course, would deny that when Darwin came at the end of the 1850s, he lent increasing plausibility to certain of the aspects of laissez-faire, giving it a more plausible, more doctrinaire, and more bigoted expression. But I think the social values were there already.

Take another, later, and not unconnected instance. In the mid-1970s we are seeing vehement, not to say acrimonious, disputes between hereditarians and environmentalists with respect to the malleability of IQs and the general tractability of the aggressive instincts and other instinctual equipment that are said to come down to us as a part of our sociobiological heritage. Well, I believe that anybody who looks into these debates, as I have recently been doing, will conclude that the environmentalists are people who already have a strong commitment to social activism and social reform and that that has, in considerable degree--I don't say wholly--led them to take the environmentalist side of those debates.

And, conversely, I do believe that people who are on the hereditarian side are people of a basically conservative posture socially and politically. My general thought is that social values have enjoyed a substantial independence (not a total independence, but a substantial independence) from scientific developments. I don't think that this wide autonomy of social values is likely to be altered in the years immediately ahead.

I add, however, one qualification and one paradox. Undoubtedly, some scientific developments have aroused such anxiety about their own immediate practical consequences for the human race as to touch off concern for social values that are thought to make it less likely that science will do all the harm that people sometimes fear it might. It is, by definition, scientific development that got us excited about radioactive fallout and the possibility of contamination of the atmosphere by the residues of chemical insecticides.

So, you do have, I think beyond any doubt, that category of scientific development that makes people desperately afraid of science itself, and the social anxieties (necessarily mingled with social values) that arise in that context do fit the pattern of science coming first and the social values following thereafter. But I still regard that pattern as a special category; however, it is likely to have more and more items in it in the years to come.

* * *

I don't think that social values necessarily produce scientific theories and developments under their genuine scientific aspect. In the case of Darwinism, one knows that there was a large input of current social theory coming from Malthus that amounted to the philosophy of the political economist.

But that is not, in my view, the general pattern in the history of science, and I am perfectly prepared to believe that most scientific developments occur because of the ongoing stream of research within the disciplines in question. I am simply discussing the issue of what happens when people decide to put what I regard as genuine scientific development to prescriptive use. What happens when you convert Darwinism into social Darwinism? I think you then find that people invoke it on behalf of whatever social values they already have.

However, there are other cases, such as the debate now about the malleability of IQs, in which there is so much uncertainty that nobody can be blamed for the side of the debate that he finds humanly more congenial. But I basically agree that derivative science has much autonomy.

* * *

There is the possibility of a synthetic science or a synthetic field of study; one that has been mentioned repeatedly, though not enough to satisfy everyone, is ecology. It does constitute a precise meeting place of physical sciences, biological sciences, and even social sciences and ethical concerns. So maybe we ought to be trying to define some new synthetic sciences that won't be fake and in which people can focus on problems and bring different kinds of science to bear on those problems. That, I suppose, isn't entirely utopian because some people are being so trained.

* * *

I don't think that social values are transformed by discovering that you could prevent people from contracting smallpox. I don't know of any evidence that people wanted smallpox before vaccination or that positive value was attached to it, and I never read an apology for mine accidents before you had the means of preventing them. The social values remained the same. Of course, science has enabled us to act upon some values that we couldn't act upon before, and that is a tremendous change in human history, on which I am sure we are agreed. But the social values didn't change at all.

ROBERT H. DICKE

Albert Einstein Professor of Science, Department of Physics, Princeton University

In thinking about modern physical theories and the effect these theories might have on social values, I am struck by one particular point. Newtonian mechanics, the great theoretical achievement of the seventeenth and eighteenth centuries, was framed in such a way that the basic elements of the theory could be directly comprehended in terms of one's own personal perceptual world. Newtonian mechanics dealt with the prosaic concepts of force, motion, and position; one could feel comfortable about an explanation based on such familiar concepts. Even the internal motions of the enormous solar system could be explained in familiar terms.

The great achievements of twentieth-century theoretical physics have involved concepts that are not so directly carried over into one's own personal world. Consider the relativity of Einstein. Here the basic idea is one of a four-dimensional space. General relativity, Einstein's gravitational theory, is based on a curved four-dimensional space, whereas the space of our perceptual world is three-dimensional, with time as a separate one-dimensional space. We cannot, therefore, very comfortably incorporate the discordant concepts of relativity into our everyday perceptual world.

Quantum mechanics provides even more extreme examples of incongruous concepts. The physical concepts of this theory seem foreign indeed to our everyday ways of looking at things. How does this come about? Why is our inner perceptual world inadequate for modern physical theory? The world of quantum mechanics is the world of the very small, e.g., the atom. But our personal, internal view of the world is based on information derived from a larger world by our eyes, our ears, and our fingertips. Our perceptions are derived from this body of sense information.

But these senses are too insensitive to tell us about the basic structure of the world of the atom. More than the sense data are missing. Concepts derived from the inadequate sense data are misleading. When first encountered, relativity and quantum mechanics seem strange. It is as though we have seen an optical illusion for the first time, that we are viewing a strange world.

One of the most interesting features of the quantum mechanics is the role that probability plays. One does not say that a particle is first *here* moving with a velocity that takes it *there* at a later time. One deals with probabilities, a probability amplitude giving the probability of a particle being found at a given point. We cannot at the same time ascribe both a position and a velocity to a particle.

I don't know to what extent the determinism of Newtonian mechanics influenced our present ways of looking at the world. For Newton a particle existed at a point with a definite velocity. A probabilistic world is a foreign concept, but perhaps definiteness is simply a characteristic of the human brain itself, a binary digital switching mechanism yielding yes-no answers. Things are either black or white. Something is either true or false. But the world of the very small isn't that way. According to quantum mechanics some questions do not have yes-no answers.

A yes-no, good-evil, black-white, true-false view of the world has permeated every aspect of our lives. This may have led to questionable legislation in the realm of environmental protection. It is doubtful that cyclamates would have been banned if federal law had permitted a balanced assessment of the probable benefits and risks involved in their use. It is possible that far more lives are lost by denying this sugar substitute to the population as a whole (and particularly to the incipient diabetics) than could conceivably be lost through cyclamate-induced cancer. It is also possible that the original experiments that led to the ban were inconclusive, or wrongly interpreted, and that cyclamates do not induce cancer. But federal law made no provision for doubt: cyclamates were either good or evil; if evil they were to be banned.

A probabilistic world-view could drastically affect societal values. (One immediate effect might be the elimination of racetracks and state lotteries.) The risk of death by traffic accident might be compared realistically with environmental risks associated with travel. The human brain is capable of reasoning probabilistically, but the end result for a system of values may be difficult to predict.

* * *

I think we all agree that this is a very strange situation that we have gotten ourselves into. But nonetheless, as you know, the attempts to find hidden variables in quantum mechanics have not been successful and probability may be with us from now on. Except for the mathematicians who prove things, scientists who construct theories (and even those who do experiments) can never be certain that their conclusions are correct.

We must live in the world that we have. Our information is incomplete and inaccurate. The question is: What do we do when we have less than perfect and complete information? Here probability is forced upon us.

I would like to respond to the matter of the probability of incorrect results. There must be a wide range in the validity of scientific conclusions we reach. But I think we can agree, if we look at the history, that the physical concepts and the theories of the past have had to be modified. There is a lack of permanence in the scientific view.

* * *

It seems to me that science was forced into a state of fragmentation by the necessity to work in a limited area in great depth in order to make progress. There is a need for scientists who do not develop new science but simply communicate and tie the pieces together. I don't think that we are training that type of person in the universities now.

* * *

I see two sources of difficulty in incorporating the more modern physical theoretical ideas into our everyday ways of thinking. One of these is that the newer physical concepts do not fit very well into our perceptual world. Each one of us carries around his own personal world, and its pieces don't fit into the perceptual puzzle very well. The other thing is a quite different matter. It is a matter of language. Here, I think, something can be done. We can devise a language to communicate with those outside of our own scientific field. We must communicate better.

DUDLEY SHAPERE

Professor of Philosophy, Member of the Committee on the History and Philosophy of Science, University of Maryland

The Copernican revolution is often cited as having led to a breakdown of man's conception of his special place in nature and of all the values associated with that idea. Today we are witnessing upheavals that can be, and often are, attributed to Freudianism and its offshoots.

The possibilities of computerization, genetic control, and the prolongation of lives of hopeless invalids are today opening up new questions about the relation of science and social values. Again, it might be argued that scientific method as well as scientific theories, structures, and results have affected social values.

Now, partly, the question of whether science has influenced social values is, or can easily degenerate into, just an argument about words, depending on how broadly one defines science--and today under the influence of the history of science as it has been developed in the last several decades, we have become increasingly aware that we cannot define science narrowly. Depending on how broadly we define this word, the views of Aristotle about matter, motion, the elements, and the heavens exercised a profound influence upon, or at least were intertwined with, social values. And depending on how broadly one defines social values, it can be argued that, say, the scientific work of Lavoisier, of his successors, or of Faraday and Maxwell ultimately had a pervasive effect on society and social values.

Under the impact of chemistry and electromagnetic theory, we do many things automatically that, if we were asked, we would probably say that we ought to do or at least that doing them makes life easier, better, or more pleasant. I assume this is what we mean by social values. In this broad sense, whenever we turn on an electric light rather than light a candle, we are expressing a social value. Material results of science then become themselves expressions, in many cases, of social values.

Scientific ideas have often been at least awakened--whether justified or not is another question--by social values. But the point is frequently overdone, carried to an extreme. Perhaps this is, itself, partly the product of a social phenomenon, namely, of the antiscientism that has become so powerful today, an antiscientism that has manifested itself as an attempt to downgrade science by seeing it not as possessing an objective, value-neutral method by which well-grounded beliefs are arrived at independently of any values, social or otherwise, but rather to see science as relative to the values of a particular culture or a particular tradition.

In part, also, the overdoing of this point is a reaction, and I would say an overreaction, against the view that science in its majestic neutrality develops wholly apart from its cultural context. It seems to me that it teeters dangerously on the brink of a complete relativism of knowledge, in particular, scientific knowledge, and I think this is a tendency that we have to resist very strongly.

* * *

Thinking about values is not thinking in the same way as is thinking about scientific questions. There have been people, of whom John Dewey is the most conspicuous example, who have held that being a good scientist defined an attitude toward the world--humility in the face of experience, an actual solicitation of unanticipated results, relishing the shock of discovery, being willing to give up positions held in the face of evidence to the contrary; there was that faith which, God knows, was vigorous before 1914. Dewey himself maintained it until 1939. I suppose it is in a defunct phase now, but I am not entirely sure because certainly if you look, say, at the Soviet Union, there are some scientists who argue that in order for them to get on honestly with their own business and arrive at the conclusions they think are justified and publish these to the world, whether that be in genetics or some other field, they need a society which is quite differently organized from any dictatorial state.

So, maybe there is that residue of sense in what Dewey said, although I come back to the view that on the whole that is an exploded faith, but not quite.

* * *

In discussing science and values we come to some very difficult questions. Now, these questions are of several kinds. For instance, there is the problem of information gathering arising out of the vast amount of fragmentation and specialization of science and the idea that cross-disciplinary efforts are necessary to deal with problems of society. How is this information going to be brought together? There is also the question of assigning priorities to scientific research in the light of the funds and other resources that are available. Finally, there is the notion that, even if there were no limitations of funding and research and resources, there might be certain areas of science that ought not to be undertaken, at least under present circumstances, and this can be called the problem of developing a scientific ethics. What I think needs to be looked at is how these questions can be dealt with by trying to deal with certain aspects of society, the institutions of learning and research, science and scientists themselves, government and the society, that is, the attitudes of the public toward science.

I think there is a critical role to be played by universities. Despite the rather superficial fact that a tremendous amount of knowledge has been acquired in a tremendous number of new fields, there are in our concept of education a lot of Greek hangovers: knowledge is for its own sake, and should be so studied, totally independently of its application to society. It seems to me that to deal with the problems before us, we are going to need people who are trained in a very different way from the way people are normally trained today in science.

For example, in order to overcome the problem of fragmentation, to be able to gather the information that is scattered throughout science and also in order to develop an adequate and intelligent scientific ethics, we are going to need people who are trained in sciences, in a large number of sciences, and not by giving them the ordinary, usual elementary course in physics or in biology, but by giving them as much as possible of an updated knowledge of the sciences as they exist now, the problems that they face, including technological problems associated with them.

These people would have to be trained in sociology, in history of science, in philosophy of science and political science and ethics and economics, possibly also in certain aspects of the law, in other words, trained in such a way that they can approach these problems in a cross-disiplinary way.

FORUM II

The Citizen and the Expert

We are aware of the tensions between the specialist, the expert, and the man in the street, the external critic, the skeptic, the citizen who forms his own opinions. Where does the truth lie today? How far can we rely upon experts? Who trusts them? Do our institutions—from the three branches of the federal government to the structure of the National Academy of Sciences itself—contain room for expert opinion and citizen decisions?

DANIEL E. KOSHLAND, JR.

Professor and Chairman, Department of Biochemistry, University of California, Berkeley

A French prime minister once remarked that there are three ways that a leader can lose power: by gambling, by chasing women, and by trusting experts. This skepticism toward experts is pervasive at a time when a technological civilization depends on experts. So the first question we ask today--when do we need experts?--exposes part of the problem.

On one extreme, such as building a bridge across a river, a citizen knows he needs an expert. Common sense alone cannot design a bridge. On the other extreme, judging a beauty contest does not require a professor of art. Yet both might qualify under the dictionary definition of expert: "a person with a high degree of skill in, or knowledge of, a subject." Quite clearly that definition fails to define the nature of "a subject," and that is very important.

In the case of the bridge, there is a vast store of objective and complex knowledge beyond the training of the citizen. In the judging of beauty, the background knowledge is peripheral and far from objective. No outside expertise is needed. These extremes are easy to handle, but there is a vast middle ground, illustrated by nuclear reactor safety, FDA drug testing, environmental protection, malpractice suits, and so forth, in which the role of the expert is not nearly so clear.

To give illustrative examples of this middle ground, psychiatrists are considered experts by courts of law; and each day we read headlines about judges who order the accused to be examined by experts of psychiatry. Yet as citizens we expect that the prosecution's psychiatrist will give one type of testimony and the defense's psychiatrist another. I have been told that in Cleveland there is a firm that will supply experts in malpractice suits on whatever side the client wishes. These are usually referred to as "the finest experts that money can buy." Economists are usually devastating in their criticism of government policy, but when asked to predict future events they produce widely divergent solutions. These common experiences of disagreement lead the citizen to doubt the authority and sincerity of experts.

To aid in this problem I would like to suggest that we give ratings to experts in the way we rate municipal bonds. Instead of credit ratings, I would suggest we call them "credibility ratings," and the rating would be judged on the basis of objective knowledge in the field. There would be a number of ways to obtain such a rating, but one simple one would be to ask twenty independent individuals in the area the same question and compare their answers. Twenty identical answers would give a rating of 1,000; twenty different answers, a rating of 0. It would, of course, be necessary in some cases to run field trials to be sure that the experts are not unanimously wrong.

It is important in this regard to ask questions in the area of policy decision. The ability to design bridges or an SST does not establish an ability to decide on their desirability. The ability to design a nuclear reactor does not make one an expert to decide on the risk ratios to the public. And this would be true of psychiatrists and economists. They may be competent in some areas; but if they testify in criminal trials or on the federal budget, they must establish that the state of their art justifies an expert rating.

We could go further and even say that it is the duty of learned societies, or even the National Academy of Sciences, to be sure that the citizen is being fairly treated and to examine certain areas in which experts are used extensively to establish the validity of the expertise that is claimed: in other words, to be sure that behind the scientific jargon there is a corpus of objective truth. Perhaps we need an FDA of expertise to certify which experts can be "generally regarded as safe."

Failure to do this has had two unpleasant consequences: (1) it has allowed the citizen to be intimidated in areas in which the expertise is nonexistent or weak; (2) it has allowed the genuine expert to be devalued so that his advice is ignored in cases where it could be helpful.

When, then, do we use experts? The answer would seem clear: whenever there is a substantial body of objective knowledge beyond the training of an ordinary citizen. If the credibility rating of the expertise in the area is

1,000, the expert should be listened to with great respect and have a decisive role. If the credibility rating is merely 500, his opinion should be taken with a grain of salt. If it is close to 0, we should consider that there is no expertise and the pseudo-expert then has the right to express his opinion, as every citizen does, but he should carry no more weight than the ordinary citizen.

This opinion conflicts with some who believe that the expert is inevitably biased and hence should be eliminated from the decision-making process. He should, they say, be put in some minor, subservient role, speaking only when spoken to. This presumes two conclusions: that the expert has less common sense than the citizen, and that the citizen has no biases of his own. If the expert, by reason of his training, is so biased he cannot act with common sense, then an active citizen prominent in public causes is logically excluded also. This would reduce us to using intellectual eunuchs, individuals so neutral in opinion that one person has suggested to me that perhaps they are neither for nor against apathy.

This brings us to another reason for experts being used in policy-making bodies: the time factor. Most of the current decisions that are highly controversial are decided against the backdrop of urgent deadlines. The cyclamate decision, the SST decision, the Red Dye No. 2 decision are all issues in which the scientist involved would have loved to have two extra years and an extra $2 million to do the studies that might have decided the issue in an objective manner. In fact, the scientist would probably vastly prefer in general to be out of the decision-making process and to be given a question, allowed to devise experiments, and report back only when those experiments have been completed.

But in issues of vast public concern, no such leisurely time schedule is allowed. In the SST decision and the Red Dye decision, for example, the latest data are being delivered to the policy body even as the final decision is being written. Does it truly make sense, then, to exclude from these bodies those who can understand the data presented? Is the ivory tower of the scientist who would like a few years to finish his experiments any more unrealistic than the ivory tower of the citizen who wants all the scientific conclusions finished in a one-page, true-false memo in time for a lay board to make the final decision?

Finally, in order to be a little controversial, I might suggest that the expert may in many cases have advantages on such citizen policy-making boards for one simple reason: he has learned to live with complexity. It is nice to believe that the issues are simply whether cyclamates are good or bad, whether genetic engineering is moral or immoral, whether nuclear reactors are safe or unsafe. In fact, the issues of greatest interest are never that clear. They are partly good and partly bad, and only the courage to face the honest fact of partial truth will carry us through. The expert must learn to be more humble at times, to be more rigorous in defining those areas in which expertise exists and in which in fact it does not. The citizen will have to learn to use the expert, for he is the one with the flashlight in the dark forest of our current technological complexities. And in the absence of absolute certainty, the public is going to have to learn not to have villains and heroes if decisions made with inadequate scientific evidence are later changed.

* * *

Some scientists have a little concern about turning everything into an adversary procedure, and not just because they are unwilling to be in a fight; I think a lot of them enjoy combat. But just as the scientist has to be educated to learn about the public process, the public has to learn about the scientific process. De Tocqueville said many years ago that the public will always believe a simple lie in preference to a complicated truth, and simplicity is more persuasive in an adversary process. Maybe the SST hearings are an example of a good adversary process. Each of us can express our annoyance at sonic boom, so the citizen needs no experts. On the other hand, in the case of cyclamates, a preliminary report (which all the scientists know is very tentative) is published in which we are told of rats being fed thousands of times the amount of cyclamates per body weight that any human will ever consume and a few of them getting cancer. Suddenly the Commissioner is pushed by the public excitement to make a decision. Maybe the courageous decision is to say that we just don't have the data, and, in spite of this one little report, we shouldn't make a decision until the data are available.

STEWART BRAND

Editcr, Whole Earth Catalog, Whole Earth Epilog, and The CoEvolution Quarterly

I would like to discuss when we don't need experts. This is pretty much respective of the citizen, and that is the area I am most involved in serving.

I think an expert is not needed or usable when we can't afford one. If we can't get the best expert that money can buy, then money is not part of the equation. It is usually the case that you can get experts rather cheaply--not in person, but by their writings or by talks or other forms, and it is the other forms that particularly interest me. It is fairly rare that an area of concern is moving so quickly that you can't find something very well thought out and very well written on it, or that you can't make a few direct telephone calls, each of which leads to a further level of expertise. In the course of an afternoon you can find out everything you need to know, starting with the yellow pages.

There is another area that is especially tricky, and that is where the actual locus of expertise is different than it appears, such as, for example, in the area of birth control pills. In recent years such expertise was supposed to reside with the gynecologist. What women began to find was that they were receiving very poor information from their gynecologists and getting much better information from other women. Anecdotal reports were more useful to women than the advice of an expert in deciding if they wanted to stay with these problematic devices.

It is fascinating to me to observe the actual ways in which expertise is received. It is not particularly scientific, though that is certainly one increment. You look for someone who is rather good at appraising with some objectivity. But then you have got to appraise the appraiser, and the way this happens is quite informal. It has to do with experience of the people whose expertise you are testing, with gossip networks, what you have heard about so-and-so and from other so-and-so's whose expertise you are also guessing at. But by and by a feeling for where the real information is develops. This is not supposed to be the case, but it is, and I think perhaps we could improve our sense of how to actually find out where the quality is.

So all this comes back to things that are not so quantifiable, such as character and integrity and the networks of trust that are built up among groups of people who know each other. They don't necessarily agree, but they trust each other. When you get into one of these, the quality of information you receive starts to improve drastically. Perhaps that is the function of things like this Forum, that there is a certain amount of face-to-face contact through which this kind of information passes. It is not quantifiable, it doesn't pass very well through the printed media, but it is absolutely essential to judge real expertise.

* * *

I can't resist one comment about developing your own expertise as a citizen. We carried a great book called the *Merck Manual* in the *Whole Earth Catalog,* first edition. The *Manual* is essentially a layman's presentation of the full body of general-practitioner medical knowledge. With a little application and experience it is quite useful in the home. So we carried the book and also made it available through our Whole Earth Truck Store. I immediately received a quite curt letter from the publishers, saying that we could not make this book available to the general public and that they would do everything they could to prevent us from doing so. We then had to say, "You can't get the book from us, but just go to the nearest university bookstore and pretend you are a medical student and you can get the book." It seems too bad that you have to go through that labyrinthine approach to obtain information that should be available to everyone.

* * *

There is a sequence that goes on in this access to and evaluation of information. You start out with access to the services that are available in the community, and as a certain confidence grows around that, then there is also a market for evaluation. For example, in a good many colleges there is usually a quite outlawed and informal publication of some sort by the students to evaluate teachers, often quite independent of the particular subjects the teachers are teaching that semester or quarter. It is often the most important guide to the best use of that college or university. It may be a fuzzy Xerox copy, but it is read and valued, and people will buy it.

The same thing can be done on the community, neighborhood, or city level--a separate evaluation document going on in quite personal and often insulting terms about the professionals that are available, indicating which are the good ones. This is a thing that the market can satisfy.

* * *

I could add a report I have heard about the use of committees and truth. I understand there was a study done on the Delphi technique, which is a means of predicting the future in a particular, say, technical area like computer science by polling a number of experts in that area to predict what is going to happen in the next ten years. There was one study that was done where there were enough experts to do a double polling. They asked one set to come together and confer for several days and come up with a set of their expectations of what was going to happen in this area for the next decade. Another equal set of experts who knew more about it than the first set were contacted by letter and telephone and asked to send in their individual opinions about what was going to happen in the future. This was then put in a file. The dramatic opening of the file ten years later found that the people who had been individually polled and not brought together were far more accurate.

FREDERICK C. ROBBINS

Dean, School of Medicine, Case Western Reserve University

It is important to realize that no one is an expert in everything and that each of us is a citizen in respect to those areas outside our special expertise. Whether scientists are experts beyond their immediate field depends not so much on the role of the scientist as it does upon their personal qualities. It may be true that a well-educated person is a better judge of certain things than a not-so-well-educated person, but Jefferson and others certainly would have argued with this statement.

* * *

Another matter about which I feel rather strongly is what I think of as "the halo effect," a condition from which persons in certain positions suffer. Individuals recognized as authorities in a particular field frequently are expected to be knowledgeable and wise about everything. I think that is something of which we must be very cautious.

* * *

I find discussion about science as an art form very convincing and very real. On the other hand, it is an art form that has some differences from other art forms, and I guess that is what a lot of the problem is about. In the conduct of one's artistic efforts in the field of science, it is possible to accomplish great good. It is also possible to stumble on something quite unknowingly that can be very destructive. I do not happen to think that the recombinant DNA research is going to destroy the world. But it is of enough worry so that the scientists themselves sat down and have attempted to lay out reasonable guidelines to control the likelihood of something harmful coming about. This is the problem in the eyes of many of the lay, nonscientific citizens who are very concerned: Can they trust the scientist as an artist? Or are these not dangerous enough toys that are being played with so that perhaps some means of participation by society in decision making is desirable? I find it quite difficult to try to figure out how this might occur without destroying the spark of scientific research.

On the other hand, I do not think those of us who promote science and are interested in science can afford to ignore this public concern. It is real and must be attended to. It may be that one frontier that science is facing is the coupling of what we have thought of as science with what we have not thought of as science, but as still important elements of civilization and society. But the worry is whether or not man is capable of dealing with what we are presenting to him. By our intellectual achievements we have exceeded our capacity to evolve rapidly enough in other ways. This is not a new thought. It has been proposed by others, including Dubos, that we are burdened with our physiological capacity to evolve in an evolutionary way. This is very, very slow, affecting intellect and emotions at almost the hunter and gatherer level of responses.

I have been impressed with how difficult it is even for highly intelligent, motivated people, when the chips are down, to think in the long range. They tend to think on the basis of short-term gain, personal gain. This is not a very suitable frame of reference for a society that has to plan over the long view. Clearly normal processes of evolution cannot deal with this. We are going to have to find other ways to deal with it, and possibly by intellectual achievement we can do it.

HANS A. BETHE

Professor, Laboratory of Nuclear Studies, Cornell University

The first time I was asked to function as an expert was right after World War II, when nuclear weapons had been invented and brought into our arsenal. My expertise in this case was not so much needed for the way in which you make nuclear weapons--that is not very difficult, and in fact, you can write it down. The difficult problem was what to do with them. The experts--the people who had worked with nuclear weapons during the war and had lived with them--had much more of an idea of what nuclear weapons would mean to the world at large, what they would mean not only to the possible conduct of war, but also to international relations. As many of you know, most of the nuclear scientists at that time recommended that nuclear weapons be put under international control and not be part of the national armaments. The United States tried and failed to bring this about.

This kind of forward perspective--what a given invention might mean in the future--can be judged much more easily by the people who have worked on it and lived with it for several years than by people who come in fresh. I perfectly agree that the final decision has to lie with the elected representatives of the people and not with the expert. Nevertheless, I believe that the expert should be heard not only in the narrow field of his expertise, but also in judging the consequences that a given development is likely to have.

N. BRUCE HANNAY

Vice President, Research and Patents, Bell Laboratories

I want to address myself to another aspect of the use of experts, namely, the question of substance. What is it that we want them to tell us or to recommend or perhaps to decide? Obviously, if our only concern were with purely technical or scientific issues there would be no discussion. Our concern is clearly with the role of experts in connection with issues that ultimately involve, in addition to scientific and technical judgments, political, social, economic, or other judgments or actions. So what is it in such situations that we can legitimately seek from scientists and engineers as particular examples of experts?

The basic question we have to face is: Is it possible for experts to separate facts from opinions? Or, to put it somewhat more gracefully: Can they recognize the point at which their expertise runs out and their biases take over, and political and moral or other attitudes substitute for scientific objectivity? I think the answer is that they *can* make that separation; they often do, but unfortunately not always. One reason that they may proceed to the point of recommending actions even when the scientific or technological evidence is insufficient is that there may be a compelling need for a decision even before we have all of the facts. Someone is going to make that decision. It is tempting, perhaps, to help it along by taking the available facts to be more conclusive than they really are. The strongly held opinions of experts sometimes blur their perception of facts. Perhaps Jacques Ellul put his finger on it when he said, "Intellectuals absorb the largest amount of second-hand, unverifiable information. They feel a compelling need to have an opinion on every important question, and they consider themselves capable of judging for themselves."

There is little doubt that there has been a certain loss of credibility for scientists and engineers. Occasional excesses in statements to the public that represent opinion rather than knowledge, perhaps given through advertisements in the public press or conflicting public assertions, are part of it. But I think equally, and of particular concern to the Academy, committee reports that lack scientific objectivity or give advice on policy in areas beyond the committee's competence are at fault. I doubt that this has reached a truly critical level, but I think we have to take warning that we are losing ground on the acceptability of what we have to say, which is too bad since most of it is sound and worth listening to. As professionals I think we would find it worthwhile to institute better quality control.

There are some who despair that a committee of experts could ever deal satisfactorily with issues involving both technological and value judgments. Arthur Kantrowitz for one has proposed that we abandon the idea, substitute courtroom adversary style proceedings so that advocates of all sides of the issue can present their viewpoints like trial lawyers, and it is indeed reported recently that the Ramo Committee plans to try this. Both of the lawyers on our panel have spoken kindly of processes that are well known to and practiced by the legal fraternity.

This is an interesting idea, but I don't think we need to go that far. I think we would do better to minimize the tendency to advocate rather than to encourage it. The appealing feature of the suggestion, of course, is that then we can just accept bias, label it clearly, and balance opposing biases while deliberately encouraging them. But in fact, a well-constructed committee contains all the elements that are required, that is, people of differing viewpoints as well as a body of impartial and knowledgeable people to add objectivity. I am more impressed with the level of truth that I have heard in committees than I am with that I have heard in courtrooms.

What we really need to do is use committees more effectively, not get rid of them. The essential thing, I think, is that we must ask the committees to do only what they can reasonably do well. That is, they can assemble the scientific and technological facts as they are known; they can assess these facts; they can relate them to social or policy issues in terms of options, trade-offs, probabilities, consequences. Committees can and should assess the possible consequences of various policy options; they should state the uncertainty associated with this assessment as well as the time and cost that would be required to reduce these uncertainties. They can make recommendations when the scientific and technologic evidence is clear. I think the instructions to committees should be to do what I have just said. What the committee cannot be allowed to do is make recommendations based on a substitution of its own value judgments for scientific objectivity. It is in this murky region that I think committees go wrong, and I am immediately in sympathy when Milton Katz suggests that the Academy advisory groups sort out, recognize, and identify whether they are rendering an objective assessment or advocating a cause, and whether they are speaking as experts within their field of special knowledge and confidence or as citizens concerning a general public policy. I would add further the admonition that they should generally refrain from giving advice in the citizen mode since their credentials for this are generally lacking. If opinions must be given, which I mainly doubt, let them be separated out, clearly labeled as opinion, and be signed.

As long as we believe in free speech, we are not going to stop some of our colleagues from going to the public with their biases. But one thing we can do is excise these biases to a greater degree from reports issued by the National Research Council and the Academies. Valiant efforts are being made to strengthen the review system for our reports, but we can go further in this direction. Committees have to recognize the legitimate role of reviews and be prepared to accept constraints on their expressions of opinion. In this connection we could greatly aid committees by making entirely clear to them and to the sponsors of a particular study what the charge to the committee is, and therefore what it is not, before the committee begins to do its work. Committees are reluctant to abandon what they have written; and we should endeavor, therefore, to keep them from writing what they are not particularly qualified to do.

Even when committees do stick to science and technology I think we can find more effective ways to use them. We can surely inject into the committee processes a wider variety of inputs without necessarily adding unwieldiness. Some open meetings at which nonmembers of the committee could express their views might be generally useful in this respect, and of course these are used in many instances. More explicit reporting of dissident views, even when the majority of the committee have agreed on a particular position, would make it clearer that these views have been expressed and considered. The greater degree of openness of committee selection and records seems entirely reasonable.

In summary, it is my view that the processes for using committees are essentially good, but that they need some tightening. I adhere to the belief that experts will retain their status as experts only as long as they talk about things they know something about. Senator Muskie has been reported as desiring to meet the one-armed scientist after hearing "on the one hand and on the other hand" testimony. Unfortunately, political decisions clearly are made, they may even have to be made, before all the facts are in. But I don't think that we should succumb to the urging to pronounce judgments beyond the limit of known facts and then attribute them to science and technology or to claim expertise that we don't possess. We must continue to state probabilities and to point to gaps in the knowledge. Let the politician deal with these if he will and must, but let us not hide from him the uncertainties that are there.

* * *

There is, in some reports, a tendency on the part of committees to sort through options and decide which is the one that they would favor for reasons that are known best to them. If you go back and comb through the details, you discover that they may be arranged to lead up to that conclusion, and that is where I would say the committee has lost its objectivity. That is, having reached their conclusion, did they really try to marshall all the evidence and make it all lean the same way so that it would seem plausible as a conclusion? I would be much happier if what emerged from their report was a much more neutral statement of the various options in terms of what the consequences might be. I don't want them to omit any of the consequences. I don't want them to imply that the uncertainty is less than it is, in support of a particular conclusion that they may state. They may be correct about it, and they may feel very strongly. But I think they should be very careful to avoid that kind of bias. I think that is what the reviewer can spot because the reviewer isn't committed to the same view that the committee is after long hours put into writing it. The reviewer looks at it afresh and says: Do I see any evidence of that kind of bias in here? I think the review is what we are going to have to rely on very heavily, in the reports that, for example, the Academy produces, as a way of avoiding that kind of bias.

* * *

The kind of thing that disturbs me is the analysis that is carried to the *end* of the logical process, as far as the science and technology are concerned, and is then carried on further. I have specific cases in mind, where, let us say, a cost-benefit analysis is done to allow the kind of decision that ultimately somebody has to make. But, in fact, the costs may be a straightforward process, and limited only by the quality of the data that are available, but the benefits are very subjectives ones. The benefit to me and the benefit to you may be quite different from what is put into that analysis, yet it is used and put forth as a basis for a final recommendation. I think that has gone too far at that point. I think the committee has substituted its value judgment. Let anybody put in any benefit that he wants under those circumstances, but I don't think that the committee has any insight as to what the value of a human life is, or any other purely subjective benefit.

RICHARD W. JENCKS

Vice President, Columbia Broadcasting System, Washington, D.C.

It seems to me not so much a question of when do we need experts--there is hardly any situation in which a task cannot be done more expertly than it is being done--as when should we reject experts even when we need them.

Life is becoming increasingly vicarious in the personal sphere: parents long ago ceded their authority to Dr. Spock, young lovers to Dr. Comfort, cooks to Julia Child. Sometimes it seems to me that the motto of contemporary life should be--"Mama, I want to do it myself"--the ultimate cry against the expert.

And certainly in the political sphere, decision making should not be vicarious. The SST may be found by experts to not disturb the ozone layer, but political representatives ought to be free to ban it anyway if they wish. Fluoridation might be found by experts to be wholly beneficial, but a city council ought to be free to decide and to reflect the feeling of the citizenry that water should not be tampered with. Educational experts may feel they know exactly what kind of textbooks would benefit the child, but an elected city council should be free to reflect the deep concerns and felt necessities of parents of children.

Lincoln once had to cope with an expert general, General McClellan, who insisted on advising him on the overall conduct of the war. Lincoln said he felt like the rider whose horse caught its hoof in the stirrup. The rider said to his horse: "If you are going to get on, I am going to get off." In short, the expert should be servant and not master.

* * *

I think we are all in agreement that we need to improve the transmission of expertise, particularly at the government decision-making level, because our lives and our fortunes depend upon it. I think, in this discussion of how to improve it, that there is a tension between two ideas. One idea is to refine expertise and validate and legitimize experts to a greater degree. But the danger of this is that it creates the possibility of forming a closed system, an establishment, an institutionalization of truth. And in this country we have always striven to avoid the creation or the sanction of any form of official truth, the more so, of course, where scientific inquiry is involved.

It does seem to me that there ought to be a greater scrupulousness in confining the expert to the field of his expertise. As a citizen, if he feels and cares deeply--as we hope he does--he can find another forum in which to express his views on the value judgment, perhaps, for example, before a congressional committee, where he will be welcome. But he should not confuse the forum of his expertise with the forum of his political views.

It does seem to me that the media and others, including the schools, should better ground the public with an attitude toward science and information that will be responsible and that will be skeptical. Finally, I think a certain chaos and confusion is better than too much structure, too much refinement in this area. Our strength as a nation, including in the scientific field, has been due to our diversity and to our contentiousness. The best protection for us in the long run is a reasoned skepticism and the improvement and opening of our procedures.

CHARLES R. HALPERN

Executive Director, Council for Public Interest Law

I have to start with a confession: I am a lawyer, and a member of that most suspect of classes, namely, Washington lawyers. If there is suspicion of expertise, there is a special suspicion of that expertise that lawyers have, and of Washington lawyers most of all.

My career as a Washington lawyer has been roughly split in half--half in a large corporate law firm and half in the field known as public-interest law, which means providing representation to unrepresented groups, be they poor people, racial minorities, environmentalists, consumers, or the like. I start with this autobiographical information because my experience in Washington law practice has very much shaped my thoughts about the way citizens get access to experts and develop the working relationships with experts that they need to affect public policy decisions in which they have an interest and that have a large technical or scientific component.

These are the kinds of decisions that are made every day by the government, frequently in administrative agencies, in the executive branch, and in Congress. They relate to our regulation of cigarette advertising, of the pesticide industry, of auto safety, of cable television, and food additives. In all these areas there are issues of public policy of direct concern to large numbers of citizens, and these questions are ones which have a high technical or scientific component.

While I certainly agree about the importance of citizens' trying to understand and educate themselves on expert matters, nonetheless I feel strongly that the citizen who wants to affect decisions made by government where there is a large technical or scientific component is going to have to have effective access to experts--experts not of the type that we just call up on the telephone for an off-the-cuff conversation, but experts who will really lend to the citizen group the benefits of their information and knowledge. It is my thesis that citizens at the present time have grossly inadequate access to experts, particularly when contrasted to the kind of expertise that corporations can marshall in order to persuade decision makers of their positions. This disparity is growing and should be of concern to the scientific community and to the larger community of citizens.

Let me take as an example of this thesis the decision on whether to permit construction of a pipeline across the state of Alaska from the North Slope to Valdez in order to remove the oil that had been discovered there some years ago. A consortium of oil companies that leased those oil fields, for good and sufficient commercial reasons, decided that, from their corporate standpoint, the pipeline running to Valdez was the proper method for taking this oil out, a basically economic decision informed to a large extent by their in-house experts who told them what was technologically feasible and what the cost factors would be. On the other side you had a number of environmental groups who viewed the pipeline route as a threat, to put it mildly, to the last great wilderness area on the North American continent. Their decision was less well informed by expert opinion and more thoroughly related to value judgments; these were certainly not people who were unbiased or disinterested. They were dedicated to wilderness values, and they saw the pipeline project as antithetical to the values that they held closest.

The Secretary of Interior had to make a decision. Was he going to issue a permit for the construction of this pipeline across federal lands, or was he not? But what interests me is the process by which the oil companies and the citizen groups, the environmentalists, built their cases. First of all, the basic fact is that they had to rely on experts. In part, their reliance on experts was a matter of persuasion. They had to persuade the Secretary of Interior and the American people that they knew what they were talking about. In addition, in refining their positions, experts also played a critical role.

Take a look at the structural advantages that the corporations enjoyed in obtaining expert assistance and advice. The first and foremost is money. The corporations were not looking to experts who would contribute their time, who would have nothing pecuniary to gain from their work. I do not mean to suggest that scientists are venal or will sell their souls; like other people, they have financial needs that they can legitimately satisfy by selling their services so long as their sale of service is consistent with their personal integrity and their professional commitments.

Nonetheless, money is an important factor. And it is not just the consulting fees in this particular case. There is a promise of repeat business for experts consulting with oil companies, and there is a network of established relationships between whole academic departments and the industries to which they feed their graduates. This is most obvious if you are looking at a discipline like geology or seismology, where there are very close relationships between the industries and the academic departments that are extremely complex and difficult to sort out. It makes it easy for the corporations to find geologists who are expert on this sort of matter and makes it relatively hard for the citizen group to find volunteer experts--not impossible, but extremely difficult. I only want to point out the fashion in which the system weights the balance. It is not a black or white situation, but one of weighted balances.

There is another important point that flows from the corporation's economic advantages, and that is the ability to choose among experts. This was a process in which I participated as a private lawyer. If you are trying to persuade an administrative agency or an executive department of your position, you don't have to take the advice of the first geologist to which you go. You can shop around and find that geologist who is both solid on the merits of the issue you are trying to address and also has other skills that you think will help to persuade the final decision makers. Again, you are not asking any expert to bend his views to the corporate position, you are simply looking for the expert whose advice you like best.

Finally, I would like to mention the fact that corporations have vast in-house expertise. They are not going out as lay people to look for the right geologist or plant biologist to present their case. They have those people on their staff who can help in the selection process--a very significant factor.

I dwell on the advantages of the corporation because, in my view, an effective adversarial process leads to the best policy decisions. The Secretary of Interior would be likely to make the best decision on the pipeline if he had the scientific facts marshalled by the corporations on the one hand and the scientific facts marshalled by the environmentalists on the other. He could compare them and encourage mutual criticism by the experts involved in the process. The problem is that for the citizen group it is very difficult to make a presentation of the same scale and sophistication as the corporations can. Typically, they lack money; they lack access to experts and the sophistication to judge among them. In terms of rewards within the scientific community, the scientist who aligns himself with citizen groups loses respectability and legitimacy; he becomes a "scientist advocate," which, as I understand the hierarchy of values within the scientific community, is a very bad thing indeed.

I believe that access to experts is critical to citizens if they are going to affect policy decisions that have scientific or technical components. There is, in the situation I have described, an important challenge to scientists and to institutions within the scientific community. An institution like the National Academy of Sciences, for example, should make it a high priority to try to improve the interconnections between interested citizen groups and relevant scientists. It is an extremely difficult process, but not impossible. The recent hearings held on the Concorde by the Secretary of Transportation are a rather good model of the process by which policy decisions with scientific components should be made. In that case you had highly sophisticated scientific experts on both sides. They cross-examined each other. The ultimate decision was left to a public servant who had a free hand to decide on the merits, and he has been presented with expert information in a manner with which a lay person can grasp and deal.

The challenge is not only to scientists here, as my remarks have suggested. I consider the citizen group that wants to affect policy decisions to be in the hands of a number of experts, and they are not all scientists. One of the most crucial experts for the citizen group is the lawyer, and up until the present time--and to a large extent now--lawyers are available only for corporations. The misallocation of legal resources, which is very similar to the misallocation of scientific resources, exacerbates the problem. So you have highly priced, highly effective lawyers representing corporate interests and frequently no one at all representing citizen interests.

In short, it seems to me that if citizens are to affect major issues of public policy they are going to have to have better access to experts of many kinds--scientists, lawyers, and others.

* * *

Frequently, the people inside a government agency view themselves as having a judgelike role, balancing and evaluating the evidence that comes into them. And their situation is frequently similar to that of a judge who has only an advocate on one side. Suppose the prosecutor comes into a criminal case with a staff of psychiatrists to try to establish that a defendant is insane. If the defendant were in court without a lawyer and without a psychiatrist of his own, he would be in a very much handicapped position. That would be so even if the judge had his own psychiatric experts. In my view, the adversary process is crucial here. If you are going to have effective citizen participation--and the pipeline is a perfect example--it cannot just rest on citizen groups coming in and stating their values. That is a hopeless, losing situation. The effectiveness of people like Ralph Nader in many forums rests on their ability to mobilize experts to inform the positions they take and then to try to present their positions in a persuasive fashion. I am afraid the lessons of the civics class are highly deceptive, as I think anybody who is familiar with the decision-making process in Washington would attest.

* * *

In many government agencies, experts, including lawyers and scientists, frequently view the industries they regulate as a likely place for themselves to work after a period of government service. That can, needless to say, have a somewhat corrosive effect on their willingness to serve as vigorous advocates while they are inside the agencies. Again, that may not be in the ordinary civics course, but it is a fact of Washington life.

CARL DJERASSI

Professor of Chemistry, Stanford University

I wish to make an argument that is neither populist nor popular. In order to do this I would like to define the areas in which experts are needed.

First of all there are two major types of problems: prospective ones, which frankly interest me more, and retrospective ones. A retrospective one, for example, would be whether Red Dye No. 2 should be taken off the market because this and that side effect may or may not occur. A prospective one would be where expert knowledge was needed to evaluate whether work should be done on a certain topic--whether to go to the moon, build nuclear power stations, or develop a new kind of drug.

These are the two main divisions in which expert knowledge is probably needed. But there is another type of division, which I think is possibly more relevant to the question of citizen and expert or citizen versus expert, and that is that most problems can be divided into three categories: the first is a purely technical, scientific component; the second refers to societal desirability and acceptability; the third deals with political feasibility.

I would like to emphasize that almost every expert is also a citizen. It is by no means clear that every citizen is also an expert. We really need experts for the first category--the question of expert knowledge--and I am going to address myself to this question in a moment. For the second one of societal acceptability and desirability, there are frequently times when the expert can make a contribution as a citizen and not as an expert, although his expert knowledge ought to help. When it comes to the third category, the political feasibility, experts are frequently used by politicians for their own purpose, and I think that experts should be careful not to be so used. It is in that area that I am least comfortable about the role of experts. When it comes to the nonexpert citizen, it is perhaps in the third category, political feasibility, but particularly in the second one of social acceptability and desirability, that I think he or she should play the main role. It is rather important to distinguish among these different categories because one usually has to proceed in that order. In general, decisions about societal desirability or political feasibility cannot be made until some information about the technical and scientific component of the question is in hand.

One aspect of selection of experts has been bothering me for quite a while. It has been bothering me in part because I have always worn two hats at the same time. I have been a university professor and at the same time I have had something to do with industry, being involved usually with corporate management of research enterprises that were not "me-too" enterprises, but rather dealt with innovative projects. For instance, I was involved in the development of the very first oral contraceptive, which was not a negligible development at the time it was done. I now am involved in trying to design completely new methods of insect control. I would say in this case people won't question that these are societally important topics. Whether we should do such work or not is part of a different question.

The thing that has bothered me is that people use "conflict of interest" or "potential conflict of interest" and "bias" synonymously. I think this is grossly unfair. I find that in general someone who has anything to do with industry is automatically suspect. Irrespective of whether industry selects that expert, or whether he comes from industry, he is suspected and much of the evidence is automatically discounted. This is so because he or she presumably is biased and subject to conflict of interest. I think the bias of the government employees, the academic, or the consumer is as great as the bias of the expert that comes from industry. Furthermore, the definition of conflict of interest that lawyers usually use is that of stock ownership. I am not that uncomfortable with such ownership provided it is fully published and everyone knows of it ahead of time. I am much more concerned about the conflict of interest that is not necessarily illegal and of which many, particularly the press and lawyers, as well as the entire legislative branch, are guilty. I am referring to the conflict of interest generated by position in society or a profession that may lead to conclusions or actions that are as inappropriate or as full of implied or actual conflict of interest as those of the person owning stock in a corporation.

We live in a society in which almost all the practical things that we do in which expert knowledge is needed in one way or another are implemented by industry. This is particularly true of the sort of industry that is controlled by regulatory agencies such as the Food and Drug Administration or the Environmental Protection Agency. I am a firm believer in the importance of these regulatory agencies. However, I am not so happy about the manner in which some of the decisions are made, in part because of the selection of experts on certain advisory committees. The selection is always based on having only people who presumably have no conflict of interest or bias on a given topic. I think that this is preposterous. This is like selecting exclusively a group of virgins, of nuns or monks, and of seventy-year-old cardinals to write a book on the joys of sex. They could write something very academic and theoretical, but for such a book you also need people who have been in bed with someone. I would invite the entire range from prostitutes to homosexuals to happily monogamous husbands and wives to tell me about their experiences, and then I would make up my mind whether or not I want to enjoy sex.

This is exactly the mistake we have been making consistently in many problems of regulatory importance--we deliberately eliminate all people who have real-life, practical, operational experience in an area.

I will lastly discuss a topic needing expert inputs that bothers me the most. It is relatively easy to make decisions--and they are made all the time by the FDA and EPA--to take something off the market or to permit something to go on the market. But it is much more difficult to determine what research never gets done because of certain actions that regulatory agencies may take. I maintain that it is precisely the people who have to make operational decisions about research who could contribute most along those lines. I will give you a somewhat embarrassing personal example. A few years ago I received the National Medal of Science for my work on the first oral contraceptive. This was a couple of years after I made the predicition--while still associated with the company (Syntex) that had pioneered much of the intial oral contraceptive work--that less and less practical work would be done on fundamentally new methods of birth control unless drastic changes in government and regulatory policy were initiated.

This is a very serious problem that I have tried to bring to the attention of regulatory agencies with very little success. My predictions were largely discounted because of my connection with industry. I did have the courage of my conviction to publish this prediction five years ago, and unfortunately I proved to be about 120 percent correct.

I would also like to give you a personal example of how someone can have a potential conflict of interest and still not confuse it with one's role as a responsible citizen. I currently am advising the World Health Organization on how to develop a long-acting injectable contraceptive, which happens to be of interest to certain lesser developed countries. I am very attracted to that idea, and I have made proposals to the WHO that would be directly competitive with industry. Yet I am still a stockholder in a pharmaceutical company that is active in the contraceptive field. I happen to be very interested as a world citizen, not just as an American citizen, in the world-wide problem of fertility control, and I am making suggestions to the WHO that if implemented will be contrary to the interests of the company in which I own some stock. I consider this type of conflict of interest acceptable provided it is in the open. The WHO does know that I own stock in that company. I am convinced that the FDA would never have considered me as an adviser in that particular context in spite of my widely recognized expertise in this field.

I am not suggesting that one should get only experts who are "contaminated" in this way. I feel that all experts are contaminated, just as I feel that most citizens are also contaminated. What we need are appropriate checks and balances. But the type of exclusion that we practice in our country is quite counterproductive. The actual operational exclusion of people with industrial connections, or discounting the evidence right from the beginning if it originates from industry, is shortsighted. In the end the people who suffer the most are the citizens.

PETER BARTON HUTT

Attorney, Covington & Burling and former Chief Counsel for the Food and Drug Administration

My comments will focus on the use of outside experts by the government as part of its own decision-making process. I recognize the existence of experts within the government, but I also strongly support the use by the government of outside experts to help in the decision-making process.

Whether we like it or not, the adversary process is part of our government. It always has been and always will be as long as there are differences of opinion. I also believe that imbalance of representation is undesirable. In my judgment, one of the best ways to help solve the problem of imbalance of representation is to strengthen the government's capacity to solve issues by the use of expert independent advisory committees. I do not believe that use of an advisory committee in any way denigrates government employees who already are experts in the field or those who are kept off the advisory committees because of conflicts of interest.

The issue should not be whether the government has its own experts who can do the job, but rather whether the governmental decisions will be strengthened by the addition of outside experts--strengthened in substance, strengthened in credibility, and strengthened in public acceptance. It has been pointed out that experts bombard the government from all sides--from the industry side, the consumer side, the professional side, and others. It is necessary, obviously, for those in the government to choose between the competing experts.

I do not believe that those with conflicts of interest should be permitted to serve in the government as part of the decision-making process. This does not mean that those with conflicts of interest cannot make their views known. They can participate as advocates for their positions before the government without participating as part of the governmental decision-making process itself.

My own views, obviously, are based upon my four years with the government. My guiding principle is that a completely open government is far and away the best government, and that anyone who wishes to participate, including outside organizations like the National Academy of Sciences, must do so in a completely open atmosphere. My paper at the first Academy Forum discussed at length the obstacles to reasoned decision making in the government and my belief that the principal hope for overcoming these obstacles lies in procedural reform and innovation. I believe that an open and proper procedure is critical in relation to where the government winds up with respect to the substance on any particular issue.

Let me start briefly with the selection process, which no one has touched upon thus far. The formation of advisory committees that advise the government, including those in the National Academy of Sciences, must be on the basis of a public procedure, not a private procedure. There should be provision for public nominations to those committees. Conflicts of interest and imbalance must be looked at very carefully. Obviously one can always obtain agreement on a committee if you select people with a bias towards one particular result. Instead, they should be selected to represent a variety of backgrounds and opinions. Finally, I believe that all the relevant information on the individual members of the committee should be made public--their curriculum vitae, articles, any information on consultantships, and stock ownerships in companies that might in some way be affected by the results of their deliberations.

With regard to the specific procedures governing the operation of advisory committess, I have evolved four principles.

The first principle is that there must be adequate public notice of every meeting of the committee. Adequate notice does not mean, for example, following the current Academy requirement of placing notice only in its own newsletter. It means use of the *Federal Register* and press releases, going to consumer organizations and telling them, establishing mailing lists, and making certain that everyone who has an interest knows of every single meeting.

The second principle is that there must be an opportunity for anyone to present his views orally as well as in writing at any meeting. Obviously, there must be some reasonable degree of restriction. You could not have fifty people show up and each demand an hour at every meeting.

On the other hand, governmental agencies that have been living with this requirement for the past few years have found that it is not a problem. It is a manageable situation.

It is important, and indeed the essence of democracy, that those who make a decision for the government or participate with the government in making a decision must listen to those in the citizenry who will be affected and wish to make their views known. This adopts, of course, the adversary process, but it is much more informal than the traditional form of trialtype adversary process in the courts. It is modeled more along the line of scientific discourse, which I believe to be much more appropriate for technical issues. A court trial is simply not a practical or reasonable mechanism for resolving scientific disputes.

The third principle, and probably most controversial, is that virtually all the committee deliberations should be completely open to the public. I would allow perhaps three very limited exceptions: national security, which I confess I think would be a very limited exception and inapplicable to most scientific issues; trade secret data, which are discussed on rare occasions before advisory committees; and invasion of privacy, where one is talking exclusively about a man's reputation or something of that kind. But all other discussions of scientific issues should be open to the public from beginning to end, including the final vote of the committee on all conclusions and recommendations.

I find that secret sessions breed suspicion and distrust of government and of advisory committees. If people have something critical to say, if scientists want to criticize the conclusions of other scientists, they should learn, if they have not already learned, to say it in public and not to whisper it in private in the confines of an advisory committee meeting. If this has a chilling effect on certain experts being willing to serve the government in advisory committees, then I say good riddance to those who do not want to participate in that way. The country will be better off having people who, like other government employees, must stand up in public, say what they believe, defend their views, and not hide behind a cloak of secrecy.

The fourth principle is that all conclusions and recommendations must be given in writing, with an adequate justification and rationale in that written report. The written report should be complete in all respects, and stand on its own feet. There should be nothing that is not in that written report if, in fact, it was considered and concluded by the committee. That written report would then, of course, be subject to full scrutiny, criticism, and attack by anyone outside who does not agree with it.

The Federal Advisory Committee Act already contains some of these four principles as requirements for federal advisory committees. That statute does not go as far in some respects as I have outlined. The Federal Advisory Committee Act may at some point be amended to tighten up some of these requirements, for example, to make all the advisory committee proceedings open to the public, even the deliberative portions, and to cover the National Academy of Sciences and other prestigious organizations that currently are exempt from the Act when they advise the government pursuant to contract. I anticipate, however, that it will be some years before all these principles are adopted and fully implemented.

I have reviewed the guidelines on public access to information that were developed by the National Research Council dated September 20, 1975, in compliance with the policy that was developed by the National Academy of Sciences on April 20, 1975. I find this an extraordinarily imperceptive approach to the desire of the public to participate in decision making by the government. I believe that the Academy ought to reconsider that policy and those guidelines and that the government should take the iniative in persuading the Academy to do so. Either the Academy should follow the same requirements as government decision-making advisory committees, or else it should get out of the business of advising the government. In short, the government itself should say that the views of the Academy will not be solicited, and will not be considered except in the same way that it considers any advocate's views, unless it begins to follow these democratic principles. I do not want to single out the Academy, however, because I would say the same about any other group that purports to advise the government as part of the decision-making process rather than being regarded as an advocate for a particular view.

* * *

I have little concern about outside experts called in to advise the government including policy and value judgments in their conclusions as long as they clearly are stated as such. To ask outside experts who have delved into a particular area in enormous depth--usually far greater depth than the government can do because of its limitations--to stop at the threshhold of the decisional process is, I think, extremely unwise. It may be that the government administrator will ultimately disagree with the value judgment that is made by the advisory committee, but he ought to have the benefit of that value judgment in making his decision. In this respect I disagree with the comment that outside experts are needed primarily where there is a discernible body of technological or scientific knowledge to be weighed. I think outside experts can also be used where there are simply difficult judgments and moral values and where people can be brought in who have background that is relevant to the questions involved.

FORUM III

The Use of Knowledge:
Frontier Expansion or Inward Development

The American Frontier spread westward to the sea. Our overseas expansion has had a limited but real effect. Yet today we look inward to the problems of the city and the despoiled waterways more than to spatial boundaries to confront our deepest problems. So it may be in science itself: It is new knowledge and new techniques in new fields that mark new frontiers. Yet the Academy finds itself involved more and more with problems of the management of the established ground, and rather less in seeking new lands of knowledge. Is this right? Is it inevitable? Is it a result of a temporary imbalance? How can we judge what effort and concern we ought to spend looking outward or inward? Who can judge? Do we all stand equal before such a judgment?

ALEXANDER RICH

Sedgewick Professor of Biophysics, Massachusetts Institute of Technology

One of the important problems today is the question of how society should regard science. What is the appropriate use of science in terms of these two alternative modes of development: first, the acquisition of new knowledge, and second, the utilization of knowledge already in hand?

Western science today is very much an outgrowth of Greek science, which existed for several hundred years. The scientists of that day had no appreciation of the fact that it would have any practical value whatsoever. The impact of science on society via technology, in fact, is a product of the early Renaissance, and now it is very much with us as an accepted mode of expression.

Science and society constitute a pluralistic enterprise. It is not a simple system, and there are no simple answers. The question that we may well ask is how should such a system be run? What are the best ways in which the government and the private sector can act? And here I think it is interesting to think back 200 years to the Federalist papers, in which the foremost social thinkers in this country of that day--Hamilton, Jay, and Madison--describe ways of organizing a government. What you find there is that these men had a very skeptical and rather low opinion of human nature. People are greedy and are not to be trusted. Their design for a government was one that embodies the idea of checks and balances. Let them all have a say, but do not let any one group have a predominant say.

I think there is a message there today in terms of how science, technology, and this whole field should operate. If we attack this in a pluralistic manner, with many different groups having control of particular sectors in which they are interested, we will in fact muddle through in a way that is very much consonant with our democratic principles.

EDWARD E. DAVID, JR.

Executive Vice President, Research, Development and Planning, Gould, Inc.

Thomas Jefferson said that knowledge is power. As an inventor he had the experience--that extraordinary experience--of bringing understanding to bear for his own purposes and witnessing the profound satisfactions of coupling his mental processes to actual events and so modifying them.

Yet, I think that even Thomas Jefferson could scarcely have envisioned today's very complex system of turning knowledge to use in the developed world. Nor could he have foreseen the competition between nations, institutions, and enterprises that is a principal feature of that scene. And he could hardly have foreseen many of the ethical and moral questions and issues that the use of knowledge raises.

The vigor, scope, and size, the research and development teams with complex instrumentation, the great biomedical institutions, the national laboratories, all of this diverse activity means that it is extremely difficult to generalize, because for every broad generality there are a vast number of exceptions.

In my view the diverse nature of that enterprise is an essential to its ability to produce new knowledge and beneficial social impact. I believe the activity I am describing is perhaps the most diverse of all of mankind's efforts, and yet there is one uniform aspect that has always impressed me, namely, that the use of knowledge is quite inseparable from the creation and discovery of new knowledge and, further, that the continuing drive to make life better implies that the coalition of science and technology will continue to change the world in very unexpected ways. The world of the year 2000 is certainly not predictable, even in its broad outlines, despite the use of sophisticated forecasting techniques, because the products of the human mind are unpredictable. We can predict, however, that those mental constructs, that is, the knowledge that is created, will be brought into use if not in one country then in some other country or region.

The process through which knowledge is brought into use has been receiving a great deal of attention lately. There is a movement in progress extending what has happened over the past thirty years; namely, science and technology have been fused together so that they are in many ways indistinguishable. We are seeing now the same thing happening to the entire innovation chain, which consists of science and technology, marketing, financing, distribution, delivery, final usage, and the secondary effects of technology. This chain is not linear from one end to the other. It is a feedback system; it is highly multidimensional. This chain of activities is usually called innovation, and we are only just beginning to understand how to make the chain responsive to the larger goals of society, while at the same time preserving the spontaneity and individuality inherent in that system.

Significantly, I think, we are seeing government intervention in the innovation chain to a very high degree. This intervention is on two levels. First is stimulation of the chain, making it more active, as the Energy Research and Development Administration is attempting to do with energy innovation. And second is regulation of innovation through rules, standards, and testing requirements, such as, for example, FDA's regulation of foods and drugs. And in many ways these interventions by the federal government are antagonistic to each other and provide the base for much of the controversy and adversary proceedings that are so well publicized today. In any case, we can be certain that just as the federal government has been a major factor in bringing science and technology together so that in most places they are hardly distinguishable at all, so the government is going to have a hand in fashioning the processes of innovation.

We cannot predict the outcome, particularly the detailed outcome, of this intervention, but we can say it has the potential of affecting the very fabric of personal and institutional freedom and self-determination on which this society is built. Just as Jefferson could scarcely have envisioned today's complex system for using knowledge, so it would have been difficult for him to see this coupling between personal freedom of action and innovation. In its simplest form, this coupling is very straightforward. It is as follows: since the innovation chain will determine the life-style of the future, government regulation of that chain can determine the allowable future life-styles.

The danger, of course, is less that of a despotic government than it is of a benign uniformity imposed through regulations that, though well intended, are inappropriate to most of the challenging and intellectually advanced innovations that could come about. Rules, standards, and regulations can at best be suitable for a majority of the cases and they are likely to impede just those extraordinary initiatives that suddenly appear from knowledge newly acquired.

This larger effect of overall regulation has been called the tyranny of aggregation. This perception has generated a call among some people for an innovation impact statement before regulations and rules are promulgated.

Regulation is one dimension of government intervention and innovation. The other one is federally funded research and development aimed at producing products for the general commercial marketplace, namely invigoration of the innovation chain. And one measure of that increasing mode of activity is the $8.7 billion fiscal 1977 federal budget for civilian research and development. That is a 150 percent increase over the past ten years in real terms. The principal activities in that large budget are three: energy, health, and basic research. This $8.7 billion that the government is spending is very fast approaching what private sources themselves devote to research and development. The federal funds, in my view, are even more significant since they tend to address the longer-term research and development activities, while industry funds generally address the shorter lead-time activities. Thus federal work will tend to establish the platform from which the world of the year 2000 is going to spring.

I do not have any answer for this situation. The government is going to play a role. It cannot be turned off. Can it play this role without stifling the spontaneity of the diverse contributors who have traditionally determined the future through their thousands of personal decisions and personal interests pursued in their own way day after day? Much thought should be given to this topic for it is fundamental to maintaining the creative nature of the nation's scientific and engineering enterprise.

So far the prescriptions for benign federal interventions center on less direct funding by government and more use of indirect funding, namely incentives. The use of incentives that animate general goals without masterminding how those goals will be reached seems a promising approach.

But I think this avenue is too little explored for a full evaluation at the present time. Despite the tentative nature of the incentive proposals so far, the idea of incentives is very common in the private sector and may well have a role in the future of the innovation chain.

In conclusion, my principal points are: We cannot predict the future, but we can say that the future is going to be different from the past and that the future is going to be determined not only by science and technology but also by the entire innovation chain.

Inward development and frontier expansion cannot be separated but have to be part and parcel of the same process. Both must play a vital role, and we cannot neglect either. The most basic issue, I think, in this whole picture is the role of the federal government in controlling the innovation process through regulation on one hand and attempted stimulation on the other. Only if we can find a proper role for government can we assure the preservation of individuals and institutional freedom of action which has traditionally been the bedrock of diversity in U.S. society.

* * *

There is no doubt that we are seeing, in the United States at least and perhaps in a large part of the developed world, some changing values related to materialism. I have seen a large number of people in this country and in Western Europe who feel that their material wants are reasonably well satisfied. What they are after now is addressed more by a dimension that has to do with information and that sort of activity than it does with the consumption of material goods. And I do not think, therefore, that a turn away from materialism, which I thoroughly endorse, necessarily means that inward development is going to be the mode of the future. I do believe there is a strong tendency away from materialism and toward something else that will require frontier development.

* * *

It is certainly not true that scientists and engineers have been left alone to pursue their private interests at public expense: 95 percent of those in this country work on very definite problems assigned to them from the outside. Sometimes they bootleg a little work on their own, but most of them are very well governed. In addition to that, we have well-established processes in this society for bringing to bear the public interest on scientific and technological activities. In one case it is called the political system. In other cases it is called the economic system. And in a final case it is called the business system. I think that a pluralistic society in which there is a great deal of small-scale decision making on such matters is the best way we have at the present time of assuring a favorable outcome of science, engineering, and their use for public purposes and for benefiting the society as a whole as opposed to individuals.

MELVIN SCHWARTZ

Professor of Physics, Stanford University

High-energy physics is one of the larger fields of physics research--certainly from the point of view of the amount of funding that it has received over the years. And it has really remained kind of the super forefront, if you like, a cutting edge of investigating the really deep new physical principles. It is the line of research that runs from Newton through Einstein and Maxwell, through the foundation of quantum mechanics, through the discovery of all kinds of interesting new phenomena on a very small scale, and to the present-day investigation of things that are as crazy sounding as the concept of whether there is something called charm, for example.

This is a field of research in which the possibility of practical application in the near future is almost never there, and the motivations of people in it are about as far from practical utility as they can possibly be. That has also been true historically, for example, from the earliest days. I am sure that Einstein, when he thought of special relativity at the age of twenty-three, back at the beginning of this century, had no idea that it would end up having any practical value whatsoever. Applicability was just not a factor in asking the questions at that time, and it still is not a factor in asking the questions.

The basic reason that we ask the questions in our field, and I think it is true in very many of the other pure research fields, is that we would like to have a deep understanding of the fundamental relationships of physical objects to each other. In macroscopic classical physics you can understand the laws very simply. But then you find as you poke around a bit that it is not all so simple, that in fact there are clearly situations in which the old classical laws do not work. So you develop something new, you work ahead, and then you discover that, in turn, is an approximation. In fact, if you look a little harder you find that even some of the basic assumptions that you had were not true. For a while the world gets more complicated; and then, of course, it becomes simpler again as somebody comes up with a basic way of viewing it.

I also have worked in other areas in which the motivation has been quite different: to develop pieces of hardware that really work and do nice jobs for people. But I basically feel that you cannot mix the two phenomena. People who do pure research do it for a reason that is totally different from the kind of practical, society-oriented, application-oriented work that is done by people who do development.

There are real problems in the future of pure research. The biggest ones are associated with the sociology of the field, are associated with the fact that pure research has been over the course of the last half-century associated with the growth of universities. And, in fact, in the last twenty years the growth of research has paralleled to a large extent the fact that universities have grown, and universities have grown because the number of students has grown. But now we have reached a plateau: The number of students, in fact, is not growing; it is decreasing. This means that universities no longer have a decent foundation upon which to build a scientific establishment.

On the other hand, the needs of science have increased and, in fact, even the funding that is going into science has increased. But the possibility of effective new careers has not increased because of the combination of tenure, the fact that many people came into the field during the last ten or fifteen years--all those things have made it exceedingly difficult for bright young people to develop active careers in the pure sciences. In general it has always been true that the great discoveries in our field are done by people who are really very young, who come into the field without the sense of history, if you like, who come in and look at problems and suddenly see some new way of looking at them. You remember that Einstein was only twenty-three when he came up with relativity. And that is not a unique phenomenon.

* * *

One of the great things about this country in its development of science in the last twenty years has been the fact that it has not directed any basic research. I think that the organization of science, especially as funded, for example, by the National Science Foundation, has been one in which really outstanding, bright, creative people have been allowed to think out their own problems, to ask themselves what kinds of things are deeply interesting to not only themselves, but to the world at large. What kind of things would we like to discover? When you begin to organize science, you lose the very creative and exciting people, and to a very large extent you lose that one important element of motivation, the need for self-thought, self-fulfillment, or achievement of your own personal goals that really makes science so exciting.

* * *

To me, science is a desire for knowledge--pure and simple. It is a desire to understand why things work, what the basic laws of the universe are. It would be most unfortunate if you start thinking about supporting science only because it is going to give you a bigger piece of something or a better mousetrap. Somehow the human spirit seems to me to be just a little bit deeper, a little bit broader, a little bit more important than that. I think what it wants is a deep understanding of just exactly how it is that things get put together. And what are the rules of the game? You know, one of the most remarkable things about science is how absolutely beautifully simple everything is when you finally get to understand it, how the most complicated, the most difficult phenomena suddenly take on a sort of characteristic beautiful simplicity. It is almost like art in a sense, like music, to look at it, to see it, to study it, and then suddenly to see it all fall together. I think the day we start supporting science because we think that we are doing something for ourselves in the way of making our lives better from the point of view of material and mechanical things we will have really lost the basic spirit of what we are doing.

* * *

I am appalled at what I consider the incredible lack of understanding of just what science is compared to technology. As I look back over the last ten or fifteen years and ask what are the most interesting pieces of science that have taken place, certain things come to mind.

The most interesting thing, for example, in astronomy has been the discovery of pulsars. Suddenly, by looking out at the stars, people have seen a fantastic physical phenomenon: the notion that you can have a star made up of neutrons all bound together; the notion that all kinds of peculiar, funny things are happening out there that we never conceived of. This is somehow science. Practical use? None whatsoever. The possibility of directing people to find out whether there are pulsars? Nonsense.

In biology, for example, there is the understanding of DNA, the understanding of the basic structure of what it is that makes essentially all of life replicate itself. Now these studies were done in a very, very deep framework. They were not done by direction. They were done by people who had a fundamental desire to understand just how DNA worked. Is there some way in which you can abstract from that complicated affair some very simple ideas and come up with a notion that there is a code of some kind? Again, these are people who are self-motivated, self-directed, who had the great fortune of being able to find resources, financial and otherwise, to be able to carry forth this research.

Well, that is basically what science is all about. On the other side of it, of course, there is a whole crew of people who have taken from science those particular bits of technology that make it possible to influence the world. Far be it from me to argue about the unimportance--these are tremendously important things--but they are not to be confused with science.

PHILIP C. WHITE

Assistant Administrator for Fossil Energy, Energy Research and Development Administration

Human beings, through most of their thousands of years of history, have devoted the preponderance of their energies to filling their material wants. The major objectives of most individuals, both for their own sake and for that of their children, have been more food, more and better clothing, better shelter and transportation, and, unfortunately, sometimes better weapons. This, after all, was as true before the Industrial Revolution as it is today. However, in those days, as Charles Reich has pointed out in *The Greening of America*, there was a sort of communal approach to this endeavor. We had the customs, the mutual relationships, even the formal religion, which were entirely consonant with this major activity of improving the provision of material needs.

And following this period, the steady growth--really an exponential growth--took on sort of an explosive nature, which I think has been generally recognized as having been fueled in large measure by the application of science through planned research and development taking place in the last couple of hundred years. A major phenomenon making growth possible has clearly been the exploitation of our accumulated reserves of fossil energy. All this has led to materialism becoming almost a universal religion, with rising expectations characterizing the most recent converts.

In parallel with this push for material things, another group of people has pursued quite different goals. These people, usually very much in the minority, have focused on the philosophical aspects of the human equation, directing their attention to the meaning of life and ways to better understand and improve our human relations. The progress in this field has followed somewhat less regular and less traceable paths, perhaps advancing in somewhat more cyclical or even spasmodic fashion. We recognize certain bursts of insight that have illuminated our human awareness in past ages and in certain societies.

It is certainly true that these lines of thought have been at times critical of the materialism that has characterized so much of mankind. In the case of formal religious dogma, it has even been in direct confrontation with emergent scientific thought and material progress. There seems to be one feature that is quite clear in the last century or two: The proponents of material progress, and those who apply the scientific method to foster its advance, have not felt it necessary to turn to the soft sciences, to theology or philosophy or sociology, to ensure the continuance of material progress--which they equate with human progress. I think rather they have tended to sometimes scorn the softer sciences.

It seems to me that now we are at something of a watershed in this relationship--perhaps a real turning point. After all, many have called attention to the problems that our physical and material progress have provoked. Maybe we really are on an endangered planet, as has been claimed. Our dwindling resources, coupled with expanding population and pervasive pollution, have certainly created trends ominous in their implications for the future and even near term.

It has been interesting to note the emergence of references to the Second Law of Thermodynamics, or increasing entropy, as a measure of this trend. The Second Law describes a process that is irreversible and inexorable. It has always been with us, but perhaps its implications are only now becoming apparent beyond the circle of scientists and engineers who were brought up on it. Such disparate thinkers as Georgescu-Roegen, Barry Commoner, and Alvin Weinberg recently have called attention to its implications. They offer this as an answer to those wishful thinkers who believe that man will suddenly tap some new and unexpected sources of energy or raw materials.

And, of course, central to this law is the notion of irreversibility, processes that go in one direction only, and energy, whose measure of usefulness is the quality or level of it. Perhaps for the layman this is most clearly seen by an example of high-temperature steam, well above the boiling point. It can do effective work. It can drive an engine or bake a potato, but the same amount of heat, the same number of BTUs that is in just warm water such as we let down the bathtub drain, really cannot do much. The two are equal in energy content. And therefore, they comply with the First Law of Thermodynamics, which is the conservation of energy. But they are very different in their usefulness. It is this availability of useful energy that is steadily declining. There is really no way to reverse that decline.

One of the implications of this, of course, is that in our conservation activities we should always make do to meet our needs with the lowest level of energy that will do the job rather than with some higher level that may be more convenient or even cheaper according to our present market values.

If now, as mankind, we are facing these problems and are having serious doubts about the ability of our scientific and technological research, handmaiden of progress, to keep us on the path of improving the quality of life, where do we turn? If it is true--and it seems to me that it is--that we must focus our technological endeavor inward to create antidotes to clearly counteract the disadvantages that our material activities have brought on us, we need, at the same time, to turn to the soft sciences of philosophy, sociology, and political science to help us direct and apply this effort.

I think we have to put away our scorn, if we have it. And thus it seems reasonable that we will be turning to our colleagues in the soft sciences to help as partners as we seek the means and, even more importantly, the collective will to conserve our dwindling resources through substitution, conservation, or, perhaps the most difficult of all, personally doing without.

Only in this way can we bring about a tolerable level of continuing economic and material activity. On the political-economic side, one of the facets of such a new era is clearly the need to reassess our willingness to share more equitably the remaining unspoiled fruits that our technology has brought us with the other passengers on our spaceship. Maybe it is no longer a case of "better things for better living through chemistry," to quote an advertising slogan, but instead, better ways for better living through scientific and humanistic insights and cooperation.

We are all accustomed to the idea of sacrificing today for our children's education of tomorrow. I think most of us have personally experienced and accepted that as an almost self-evident need. What is now going to be required is to sacrifice today for the needs of our great-grandchildren and their great-grandchildren, a much more tenuous imperative.

Glenn Seaborg suggests a new society that will perhaps have to exercise a quiet, nonneurotic self-control, display a highly cooperative public spirit, and have an almost religious attitude toward environmental quality and resource conservation. Translated, that means to me clearly that the use of knowledge for our third centennial is going to have to be primarily for inward development.

* * *

The Green Revolution has generally been characterized, of course, as tremendously helpful in building a better tomorrow. But this question has been raised: Thinking philosophically about it, as we begin to apply this actual new knowledge and new capability, we must ask if it is going to build a better tomorrow. Or is it simply going to support a growth in population that is almost certainly going to ensure a worse tomorrow? It is a philosophical question that I am not sure we are wise enough to see the answer to today, but it is part of this very difficult problem of how knowledge is used.

* * *

The question of motivation for pure science is, without any doubt, a valid one. However, I think there are many to whom it does not apply fully. In years of industrial research I have encountered people who become rather concerned and frustrated if they feel the results of their work--and there has been some very good science, perhaps not quite as elegant as the high-energy physics level--do not find some useful application to mankind, some way that they are going to have some implementation for the future. I think we should be grateful for the fact that, while we have motivation for pure science, we also have many good practicing scientists strongly motivated to turn out something useful and to see the results of their work actually brought to society's benefit.

JOHN R. PIERCE

Professor of Engineering, California Institute of Technology

The thing that alarms me most about federal intervention in science is what I would characterize as mammon worship: the belief that appropriating billions of dollars will automatically, through the magic of science, produce things. Giving $5 billion to the Sierra Club would not necessarily cure the problems of the environment. I know that there is an organization that advocates giving many billions of dollars for producing fusion. While I believe we will have fusion, it is not at all clear that spending such amounts of money immediately would bring fusion any closer. It might just bring confusion.

Our whole society is based on knowledge, but there is not as much knowledge as people sometimes believe. The knowledge at the cutting edge is sometimes only a sort of knowledge in principle, or only a hint. This is what went wrong with fusion. No one doubts that if you hit two nuclei together hard enough they will fuse. Indeed, that happens in the sun. But we do not know how to hit them together hard enough, and such knowledge is very, very hard to obtain.

In solid-state physics, before the invention of the transistor made it a big business, there was a sort of general knowledge. But even when the transistor was invented, it was found out that people really did not know as much as they might have wished. A great deal of knowledge has been gained since that time because people discovered all sorts of phenomena they did not understand when they came to examine the solid state in great detail.

One thing is clear in the present, and I hope it will be clear in the future, and that is that our whole life rests on technology. This has enabled a small minority of our population to produce all the food, the minerals, the manufactured goods, the transportation, the communication, the power that we all use. That minority is about

15 percent of the population. About 1.5 percent of the population produces all the food that feeds us and a good many other people in the world. Unless our technology remains good in science, in basic knowledge, and in its application, our physical standard of living will decline in one way or another not only through environmental impact, but just through having less. In fact, signs of the decline can already be seen.

Technology serves us better in some areas than in others. The physics-electronics axis is outstanding. The good activity extends from Nobel Prizes to pocket calculators. The telephone system is really a part of this axis. The telephone system is the most complicated machine in the world, but it functions very well, and, internally if not externally, it is continually changing and advancing. Agriculture, though I do not know so much about it, is very successful. It appears, from the outside anyway, that science and technology are linked very effectively.

Automobiles and much American heavy industry are sort of midway on the scale of exploitation of science and technology. You do not see new ideas entering as fast as some of us would like, or deep knowledge of the processes that go on in automobiles.

Medicine exhibits high science and technology very effectively applied at a very early state. Indeed we are told that we sometimes apply things that we do not really know. On the other hand, in terms of medical care, when entering a hospital or a doctor's office things are often sort of disorganized. Maybe they really are not. But health care is more expensive and confused than it ought to be.

Construction is very slow and expensive. Contrast it with the whing-ding way that in electronics the Nobel Prize winners are linked to the pocket calculator--and they really are by a whole series of stages. You feel that construction is somehow slow and expensive and that we are losing, that our living quarters are getting smaller and more cramped for the money we can afford.

Cities and their services, including education, are full of high-flown words, but it is very hard to detect any linkage between knowledge and operation.

Well, where and how do science and technology succeed? Where there is a close link among discovery, invention, and development--in all the little stages that are necessary to apply knowledge, and the knowledge that has to be added to the process to make something work, and in actually producing and operating a piece of equipment? Sometimes there is a linkage. In the telephone network, operation is very important because the cost of maintenance falls upon the telephone company and the process of operation falls upon the telephone user. If you made an unusable telephone set, service would collapse.

This close linkage among all these different worlds--discovery, invention, development, production, and operation--which can really link very closely, is exemplified in the military, where the government pays money for something that it wants and is going to use; in space, where the government pays money not only for knowledge, but for equipment and for the operation of the equipment; in telephony; and generally in the high-technology field of electronics. One of the big problems is how this interrelation among all these aspects of the harvesting of knowledge can be fostered. The problem is very great.

When the national government is a participant, as it is in the military and space, where it uses the equipment and gets the results, as well as paying for the knowledge, things go best. Where the government is not a user, its problems in spending money effectively are much greater. Barriers have been overcome in agriculture because farmers are good guys and help is given unstintingly and is sort of forced on them. They are really not asked to pay for the help. There have been some isolated instances of success in NSF's RANN (Research and National Needs) program. The government must support high science for intellectual growth of man as well as for practical purposes. But the chief problem at the moment is for the government to find a way of promoting the linkage of high technology and deep knowledge to the actual operation of our society in a non-costly way.

* * *

Direction of scientific research exists very strongly in government funding, but by a process that has not been completely recognized. World War II was very strongly influenced, in our thinking at least, by such things as radar, ballistic missiles, and the atom bomb. They were spectacular. This led to funding in certain directions--things that were roughly connected with nuclear physics. If the war's outcome had been as strongly affected by psychology or biology, I think that there would have been a strong pattern of funding in another direction. People in government agencies get excited about this thing or that thing, sensibly or not. Huge funds develop. For instance, people get excited about space. Undoubtedly this has led to a lot of understanding about phenomena in the solar system and the other planets that we would not have had if it had not been for this large space funding.

You might say that this is a strange and random sort of government direction. In some ways it is not so bad, because such direction grows out of spectacular results. The science that does not show spectacular impacts on our general life is perhaps not the place to put large amounts of public funds until it does.

RICHARD B. SETLOW

Senior Scientist, Biology Department, Brookhaven National Laboratory

Most of the time I think of myself as a basic research person. In many of the biological subjects this preoccupation with basic research happens to have some very important practical payoffs.

I can exemplify one of the kinds of problems that need expansion of our frontiers by starting with a quotation from Dennis Gabor's book, *Innovations*: "The most important and urgent problems of the technology of today are no longer the satisfaction of primary needs or of archetypal wishes, but the reparations of the evils and the damages wrought by the technology of yesterday."

A more important role that science can play is not in the identification and alleviation of some of these past things, but the identification of potential future evils and future damages wrought by the present and future technologies. An expansion of our frontiers is necessary to solve this kind of future problem. There is no question that we need both, because unless we have some expansion of our basic research and general way of looking at problems, inward development is going to be the equivalent of stamp collecting.

Let me give you some examples that are in a sense aimed at society and the government, who really need an appreciation of the fact that to make predictions we have to know more. I do not know what we have to know. That is what I mean by expanding the frontier. The best example that I know of is one that I happen to have been involved in: the supersonic transport. This was the result of a new technology, and it became apparent very soon, for a number of reasons, that this new technology could result in a decrease in the amount of ozone in the stratosphere. It was possible, purely from basic research estimates, to make calculations as to the amount of decrease of ozone.

And from such pure, basic calculations--no one has ever measured it--it was possible really to make quantitative predictions about what the effects might be on man if such fleets flew. The important thing is that such predictions were made on the basis of biology theory.

The theory developed not because anyone wanted to solve the supersonic transport problem. The theory developed because a whole bunch of people were inquisitive about what happens to animals when light hits them; there were no practical motivations involved at all. The point is that there are long-range predictions of a useful nature to society that can be derived from basic research or expansion of our frontiers.

Let me give you another example. The Energy Research and Development Administration wants to make predictions for the future as to what the dangers would be of various energy-derived pollutants. They ask us to develop dose-response relations for pollutants. How many people are going to be affected for how many coal plants? There is no way we can do that. And I emphasize *no way*. We cannot do experiments on people to obtain these relations. The only way we can do this is to collect a lot of basic information--and I am not really sure of what kinds--and apply it so that we can extrapolate all the way back from nuclear physics to molecules, to animals, to man, and make a prediction. We cannot do the experiment, and many administrators and managers and politicans forget that. We cannot make the prediction without a lot more basic information than we have now about how things work.

I think some of the most important problems that we have to worry about are the interactions between the environment and us, and these are sociological as well as scientific problems. If we could solve these important problems of these interactions, we would. We cannot. We have to learn a lot more about the molecules involved before we can make these needed extrapolations, and so my point is that we really must expand our frontiers to protect ourselves.

* * *

In the pure science racket there are lots of us who really get a thrill out of basic research but who do not do physics. The point I want to make is that as far as our society is concerned, this lower-order kind of basic research is absolutely essential. The distinction that I want to make is between short-range problems and long-range problems. A short-range problem is delivery of health care. A long-range problem is whether we can live with the environment. The Sierra Club is concerned with long-range problems, but it does not know how to solve them. No one knows how to solve those problems, and the only way we are going to solve them is to let some kooky, motivated young scientists have their heads and go to work on general notions of what makes biological and ecological systems work. You cannot direct expansion of the frontier. If you do, it is not the frontier. That is sort of a definition. And it is essential, I think, that everyone recognize this. These are long-range problems that need long-range solutions, and that is what the expansion is. We can encourage it but not direct it.

* * *

Only a very small fraction of the people in the world have a lot of material goods that they are willing to give up. The great majority of people do not have material goods, and I do not think we should say that the great underdeveloped world should forego these material goods that we call advantages. And so the problem is not that we should give up things. The real problem is how is the rest of the world going to come up to the standards it aspires to without destroying the planet on which we now live? And we are not going to solve this by giving up something. We are only going to solve this by finding new solutions to the problems of the great underdeveloped world.

ROBERT W. BERLINER

Dean, School of Medicine, Yale University

The basis for much of the relatively generous public support for science relates to the fact that people consider it likely that from it will arise effective and efficient applications. So I would prefer to consider what we can expect from science rather than the motivation of the people who do it. I think it is clear that the motivation of individuals varies over a wide spectrum of mixes in various proportions for different people, including the thirst for knowledge and a desire to use knowledge. Fortunately it turns out that knowledge is often useful, so that we do not have to concern ourselves too much with the motivation of individuals.

I think it is also probably clear that the needs vary among various fields of knowledge for frontier expansion on one hand and for inward development on the other. But I would venture to speak only for one area of science, that which relates to health and to medicine.

As everybody who reads newspapers knows, there is a great deal of criticism these days about our ability to deliver medical care, which is often mislabeled health care. And although much of this criticism is justified, it really falls outside the realm of biomedical knowledge and in that of economics and social organization. So I am going to leave that aspect of the problem out of consideration and talk about what biomedical science has to concern itself with, that is, primarily with what the so-called health care system has available to deliver.

When we look at what we have available to deliver we find that it is extremely variable and spotty. Some of it is very good indeed. And some of it is very poor. If we were to try to sort out what we can offer into various categories of quality and examine the reasons for the differences, I think it might provide some guidance in attempting to answer the question before us, that is: How much effort should be placed on expanding the base of our knowledge of health and disease, and how much on applying what we already know?

Obviously the best that medicine has to offer is the total prevention of disease. In a limited number of diseases, most of which are infectious, this has been accomplished or at least is feasible if we can make available to everybody what is available to at least some. This is possible because immunization is a very highly effective technique for preventing infection with a number of different kinds of microorganisms. In turn our current success is the result of about a hundred years of accumulating knowledge about these diseases, their causes, and their transmission and a number of other things about them. And without that basic knowledge, our current success would not be possible.

Of course, at this point somebody always brings up smallpox vaccination. So I have to comment that smallpox vaccination was indeed developed many years before anything was known about viruses or even bacteria and when nothing was known about immunology. But the natural occurence and recognition of an attenuated strain of a virus that causes human disease remains a unique event, and it is not likely to be repeated. Without the knowledge of specific causes and the means to cultivate and modify them for use in vaccines, we would have been a long time waiting for the happy accident that might provide us with a vaccine for measles or smallpox or poliomyelitis.

The next most effective thing that medicine has to offer is the ability to cure promptly and efficiently. The use of antibiotics in such diseases as pneumonia, syphilis, and tuberculosis is an example so commonplace that we are beginning to forget what these diseases meant in the not so distant past.

It is no accident that the most effective measures that medicine has to offer are in the realm of infectious diseases, since microbiology was the first field in biomedical science to undergo the modern transformation. And it is not surprising that other fields have lagged behind since they really started at least a half a century or more later, and it turns out that many of these are much more complicated.

We are not without success stories in some of these newer fields. The recognition of deficiencies has made possible things like hormone replacement in endocrine disorders. Derangements of metabolism have been recognized, and ways to circumvent them have been found as is the case in such disorders as galactosemia and phenylketonuria. There are numerous other similar examples, but in general the most important thing to emphasize is that success has largely followed rather than preceded the acquisition of an understanding of the nature of the diseases and their causes and the science that underlies them.

It is true that drugs have been discovered that make it possible to control some disorders that we do not really understand. Digitalis was found to be useful in the treatment of dropsy before it was recognized that heart failure was the basis of dropsy. And quinine was found effective before we knew anything about the cause and transmission of malaria. But these were accidents like smallpox vaccination, not likely to be often repeated. Modern pharmaceutical development, although still highly empirical, is much more often on a more scientific basis. But we do have some drugs that are more or less effective in treating diseases that we do not fully understand. We can control hypertension even though we do not understand its cause, and we can cure a few varieties of cancer despite the fact that we have much to learn about the disease itself.

It is important, however, that we consider those large areas of human disease for which what we have to offer is woefully inadequate. And here we come to those medical efforts to which Lewis Thomas has given the designation "halfway technology." They in turn are giving current-day medicine many of its headaches and much of its well deserved reputation for costliness. The prototype of this category, of course, was the iron lung for the treatment of bulbar polio, a cumbersome, expensive device for dealing with the end result of a disorder we were unable to prevent, whose progress we were unable to interrupt, and whose effects we were unable to reverse. The modern-day equivalents are hemodialysis for patients whose kidneys have been destroyed by disease, the cause and progression of which we do not fully understand, and the coronary bypass operation to compensate for the blockage of coronary arteries by atherosclerosis, the cause of which is imperfectly understood and the progression of which we do not know how to interrupt.

Beyond these, a couple of the more spectacular examples of halfway technology, we come to what Lewis Thomas calls "nontechnology," what we have to offer for those conditions that are even less amenable to effective treatment than those that can be compensated for with halfway technology.

Here we find all those things for which we can supply only symptomatic relief while the disease runs its natural course.

It seems clear to me that we are able to deal really effectively with those conditions we really understand, although understanding does not necessarily assure the availability of effective management, as, for example, in the case of sickle cell disease. It is only a stroke of fortune that makes it possible to deal effectively with conditions whose cause and progression have not been clarified.

The inescapable conclusion is, then, that a significant improvement in the management of our most intractable current problems depends on expansion of knowledge at the fundamental level. Clearly, we cannot afford not to try to build applications on what we already know and that has been the practice for many years; I think, possibly, we have applied more than we really know. We do have to recognize that the foundations on which such applications are built are very weak and we should not expect much of the structure to be very durable.

* * *

We know that medical science has, in fact, produced problems. I mean, the population problems are attributable, at least partly, to improvement in longevity in the population. It certainly has very marked effects on the number of older people in the population, and on the total number of people, and these are serious problems. I think the problems related to genetic engineering, as far as modifying the human race is concerned, have been greatly exaggerated. I do not believe that that is a real threat. There are certainly some hazards involved in the kind of manipulation involved in recombinant DNA. These have been recognized and are receiving very serious consideration before people venture too far into the field. I think that this problem has been handled very responsibly. I believe that the problem can be, in fact, managed and that regulations covering this field will be issued that will make it possible to proceed without serious hazard.

HAZEL HENDERSON

Co-Director, Princeton Center for Alternative Futures, Inc.

I want to focus on those inner frontiers that I think are just as exciting as the expansionists' physical frontiers, particularly the difference between knowledge and the growth of consciousness. I am interested in expanding consciousness and expanding awareness rather than the growth of instrumental knowledge because I think that, even though it sounds idealistic, we are going to have to grow in wisdom, and I do not think we are going to do that using the route that we have been using, the instrumental route. Having created this Gordian knot for ourselves, we are going to have to understand that maybe the only way out now is in another dimension, to look up or to come up with a new paradigm. So it is the truth in other dimensions that interests me and how we bring this together with our current scientific knowledge stock.

I share the growing concern of many Americans that our science and technology have been misdirected and now create more problems than they solve. We are very good at production, but the question of distribution is before us. And that is much more difficult. A deeper problem is the "religion" of materialism and the fact that our scientific enterprise has been too much guided by the profit motive. We have to understand now that free markets were not derived from God or from any kind of natural order in human behavior. In Karl Polanyi's wonderful book written in 1944. *The Great Transformation*, he points out the great paradox that free markets and laissez-faire, far from being derived from God, were indeed a package of bitterly contested legislation in England prior to the Industrial Revolution. Indeed, basically they rested on the enclosure of land, enabling one to turn land into a commodity, and the driving off of formerly independent peasants and cottagers, in order that their labor could also be turned into a commodity. And that package of legislation, over which a revolution and war was fought in England, was entirely overlooked by Adam Smith, who took it as a set of conditions upon which he then built his theory of the "invisible hand." And I think that now we are having to understand that free markets, as Todd Laporte says, cannot be used to allocate resources or program decisions "when their consequences are socially indivisible." And that is where we are now.

I think that this impinges on all our science policy making, simply because we are now talking about the public purse and very, very few of our scientific enterprises are being conducted by private enterprise. And they all have these indivisible social consequences. Therefore, this is from whence derives the great new question of legitimacy and the fact that voters have understood quite rightly that if we are using public funds, then everybody has a say in the direction of the future technological advances.

Science has allowed itself to become captured and captivated by the high living standards and the status that have been possible and offered for pursuing the goals of powerful corporations and powerful public agencies. The goals of knowledge have become power and wealth, and so science too often has become the servant of the powerful. Books such as Phil Boffey's *The Brain Bank of America*, while uncomfortable I am sure for many scientists to read and deal with, are very important. Universities today too often are turning out intellectual mercenaries whose lances are for hire to justify whatever policies the particular institution would like to pursue. I think that we have to redirect science now toward moral goals, and particularly the goals of equity.

I do not know why this question of sharing induces such fear, because we all were taught to share when we were children, and I think that many of us sense the social costs of individual accumulation. We have to go back to our founding fathers who understood very well the difference between private property for the purposes of personal autonomy and security, and private property endlessly accumulated so that it becomes multinational corporations and can be used to abuse others. And this balance is something with which we are going to have to deal. In the crowded world that we have today, morality for the first time has become pragmatism. This is why, even though this kind of thing may sound idealistic, only this kind of idealism is going to get us out of the box that we are in.

We have to get away now from the questions of feasibility that have occupied us for the last two hundred years and what I call the "mirage of efficiency." There are some processes that cannot be rendered more efficient by the application of science and technology. It still takes nine months to make a baby, and it still takes two centuries to grow a beautiful hardwood tree. We have misdirected our resources toward the kind of "big-bang" technology at which we have been working so diligently, and we are beginning to realize now that it is not neutral. We are learning that some technology seems inherently to concentrate power and wealth and knowledge in fewer and fewer hands as the result of making the rest of us more powerless and poorer and increasing the aggregate of human ignorance. A perfect example in the intuitive reaction against nuclear energy is the understanding in a democratic society that nuclear energy is inherently a totalitarian technology. This is what people fear. They understand the dangers of this sort of elitist risk analysis where there is no possibility of democratic processes.

I think science and technology can no longer hide behind the myth of "value-free objectivity." Reality is what we pay attention to. All science is normative and culture bound, and the normative choice for all scientists is the choice of what to pay attention to, which problem to play with and to research.

So we are now in this reductionist instrumental trap, this manipulative rationality, which is not fully human. This is the revolt that we are getting now from nonscientists. Perhaps this narrow approach served a purpose in the past, because it brought us to the point of being able to see the planet whole, as it really is. Now we must develop holistic concepts. We have to synthesize the Tower of Babel of disciplines. And we have to learn that we can only understand and study whole systems as with whole people--all reintegrated. The whole mind-body split is at the bottom of this, I believe. For example, we are dealing with these idiotic dichotomies between the "public" and "private" sector that lead us to believe that we can "afford" enormous automobiles and pet foods and cosmetic industries in the private sector, but we cannot "afford" teachers and policemen and firemen and basic life-support services in the public sector.

In addition we must explore the question of property rights versus amenity rights, the way we have overrewarded competition, and why we have left cooperation as an unrewarded activity to be performed largely by women. We have overrewarded analysis, and we have punished synthesis. Part of the revolt that is going on in the political arena is against the statistical illusions that all of these inoperative concepts and little Cartesian schema for trying to manage our processes have created. People realize now that it is not only as the computer people say, "garbage in, garbage out," but it is "paradigm in, paradigm out." We are realizing now that we are drowning in bad data, inappropriately collected, using inoperative, useless, obsolete paradigms. And this is what I think is on the agenda now. We have a great deal to be humble about. Being humble makes me open and more able to learn. Therefore, humility is a very useful attitude for us right now.

We must now look at ourselves, our own motivations, our internal conflicts. Because it is now too dangerous for us to keep on objectifying our internal conflicts on to the world and each other in what I like to call the "Kilroy was here" syndrome: We just love to scratch our initials all over each other and the environment. Basically, of course, this is because we are not dealing realistically with our own death.

Death is a very important thing for us to study right now. We could learn a lot about ourselves if we realized the extent to which so many of the things that we do and the cultural games that we play are because we are trying to create meanings and to hide from the fear of nothingness and the fear of death.

Lastly, we have reached the point now where we do have to discover the operating principles of this planet and that they are, I think, sharing and humility and honesty and cooperation and peacefulness and love. We know what we must do. We have always known what we must do. All the ethicists and the great religions of the past have known these operating principles. We do not need any more research to rationalize not facing up to what we must do. I love to imagine that this planet is a Skinner box and it has all the operant conditioning necessary for us to learn what we must learn. If we do not learn this, then we do not deserve to be an interplanetary species. We would become a cancer in the solar system. And so I think that now we are facing this kind of a crunch, and we have to begin with ourselves.

* * *

I would like to focus a little bit on the matter of basic research, particularly relating it to agriculture and to the dilemma of the Third World as far as food production is concerned. I think that with the question of world food production there is another perfect example of how we get into this sort of intellectual trap of thinking that you can have a very small, sort of blitzkrieg effort to produce food for the world in one geographical area and then distribute it all around the world. It is all part of our hubristic, Lady Bountiful way of trying to deal with world problems. Many of the problems the Third World nations have are due to the configurations of prior colonialism: the fact that they were hooked to this roller coaster of world trade and cash crops and all of this kind of thing. Now we are going to go in with our high technology and use this sort of blitzkrieg approach to problem solving. I think that this is part of the inherently destabilized condition that we have in the world, that we have to move more to helping them to recover self-reliance and self-sufficiency rather than this very centralized mode of trying to provision the planet.

RICHARD R. NELSON

Professor of Economics, Yale University

The relationship that has evolved over the years between science and economic progress has been profound, complicated, and particular. If by economic progress we mean enhanced ability to meet human wants, the way economic progress has gone on has been extraordinarily unbalanced to an extent that is much too seldom recognized. The bulk of our economic progress has been concentrated in a particular set of industries and activities with many others operating today in ways very little different from the way they operated many, many years ago.

When talking about economic growth or economic progress, it seems to me you have to be careful about specifying direction and characteristics, rather than fastening on the concept of an "aggregate rate." I think the discussion about rates of economic growth has obfuscated the problem.

If you think about the past trajectory that we have taken it has become increasingly problematic. We cannot continue along that trajectory for much longer without running into resource scarcities and increasing problems stemming from the by-products of the kinds of economic progress that we have had in terms of pollution, environmental contamination, and what not.

If you ask why have we gone that way, it seems to me that there are two different classes of answers, both partially right. Part of the answer, I think, is institutional. We have grown the way we have because of the way we have organized and structured the incentives that have guided the system. But part of the answer, it seems to me, is that we have moved relatively smoothly and rapidly down those paths that science--the physical science variety--and technology related thereto have opened for us. The areas in which we have moved very slowly or hardly at all are those in which our understanding and technology have not improved very much.

I take the perspective that future economic progress along the same trajectory that we have gone in the past is both limited in potential scope and subject to sharply declining returns except in terms of improving the human condition so far as the Third World is concerned. The kind of economic progress that we want, if we can achieve it, involves quite different trajectories, quite different directions. The kind of science we need to enable us to move in these directions, I suspect, is a different kind of a science than that which has been so blooming successful in the past. We need a science that really enables us to understand human interaction, human relations, a lot better than we do. We need a science that enables us to comprehend alternative organizational arrangements and how they work much better than we do. At the root we need what the great nineteenth-century philosophers and political economists meant by the sophisticated political economy.

In this sense, clearly the kind of dialogue that has been talked about between the hard sciences and soft sciences needs to go on. The kind of research that needs to go on is not inner-directed. We need to probe new frontiers. We just do not have anything like the scientific knowledge that we have in the natural sciences in economics, political science, sociology, or psychology. And to achieve a more effectively operating system is not just a question of resolve. It is a question of real lack of knowledge of how to do a lot of these things. That, then, poses the issue that bothers me profoundly.

Why do we not have this kind of knowledge? Because the fields are inordinately hard to understand and comprehend. I think it is significant that John Von Neumann, many years ago, took a look at economics and quickly scurried out of it as a field that was too hard.

Therefore, we face a fundamental policy issue regarding what we are going to do about the scientific establishment over the next twenty or thirty years. Maybe the diminishing returns to past trajectories of economic growth imply that there are diminishing and low returns in terms of social payoff to pushing the old sciences as they have proceeded. And yet the old sciences, or many of them, have been powerful sciences. I think the social returns, if you can really improve understanding of economic, political, social, and psychological mechanisms, would be enormous. And yet no one of those sciences has proved that much of a smashing success. Where we most need the fundamental knowledge, it seems to me, we have had the least impressive performance in the past in developing it.

* * *

The past economic progress that we have had, I think, has been of a fundamentally liberating kind. But where do we go from here? How can we preserve what we have achieved here that is worthwhile preserving? Where are the really important new areas for exploration? A useful example comes from the arena of medical care and the contrast between the rapidly progressing and fruitful endeavor to understand the nature of human illnesses and our limited capabilities regarding the system for the delivery and organization of medical care where we have a great deal of difficulty describing, much less understanding, how it works.

These two different arenas interact terribly strongly. In fact in many cases our inability to understand the social or economic organization problem and to get it under some system of salutary social control can twist things in such a way that what would appear from the vantage point of the natural scientist or the engineer to be a humane advance is turned against ourselves. A good example has to do with the new kidney machines.

In principle I think we would like to take a position that a new knowledge and a new capability cannot hurt, that it can only help. But that statement only holds, it seems to me, if we have faith that the organizational structures and the processes that are going to make decisions regarding the new option are wise and salutary. In many cases that just is not the case. In some cases the issue is to get on with doing better and more effectively things that we know how to do. But in a large number of cases we do not know what to do. Our understanding of what would happen if we implemented particular policies or particularly organizational changes is fundamentally very weak.

* * *

As an economic theorist, I am indeed convinced of the essentiality in a scientific enterprise of hunting toward the salient simplification. It is that that enables us to understand what we see and to see things that we did not see before. Sometimes I worry that it enables us to see things that are not there. But that is another matter. I suspect, however, that the degree of simplification that physics--or at least some branches of physics--has been able to achieve, and to which other disciplines have looked with hope and with aspiration, is enormously greater than we are ever going to be able to achieve in the really hard sciences that are dealing with the tougher problems of what is going on out there. My conjecture is--and I say this with humility and fright--that physics has picked some relatively easy problems on which to work. The problems of the social sciences are much "harder."

* * *

The question regarding the direction of the basic research enterprise is a fundamental one, one with which the country will be wrestling for some time. It seems to me important to recognize that there are two different kinds of issues involved. One is the extent to which the government should plan and program research in a field, versus the extent to which project proposals should come from scholars and the field allowed to evolve in a pluralistic way. Here, I certainly think we must preserve a system in which scholars have a lot of initiative.

Another question relates to the criteria that should be used for allocating funds across projects and fields. One kind of criterion is concerned with potential social benefits if the projects are successful or the area advances. Another kind of criterion derives from the internal logic of the evolution of a scientific field. I think it would be a mistake to put undue weight on criteria relating to social merit. But I also do not think that we can afford to let scientific ripeness of a field be the sole criterion. There is a real tension here. We see it in the present debate about how the National Science Foundation should allocate its funds. The scientific community has operated for a long time as if it were obvious that the allocation of funds should be determined by the "Republic of Science." It simply is going to have to get used to the idea that science no longer is, and should not be, a completely self-governing system.